The Implementation Costs of Agricultural Policies

OECD

ORGANISATION FOR ECONOMIC CO-OPERATION AND DEVELOPMENT

ORGANISATION FOR ECONOMIC CO-OPERATION AND DEVELOPMENT

The OECD is a unique forum where the governments of 30 democracies work together to address the economic, social and environmental challenges of globalisation. The OECD is also at the forefront of efforts to understand and to help governments respond to new developments and concerns, such as corporate governance, the information economy and the challenges of an ageing population. The Organisation provides a setting where governments can compare policy experiences, seek answers to common problems, identify good practice and work to co-ordinate domestic and international policies.

The OECD member countries are: Australia, Austria, Belgium, Canada, the Czech Republic, Denmark, Finland, France, Germany, Greece, Hungary, Iceland, Ireland, Italy, Japan, Korea, Luxembourg, Mexico, the Netherlands, New Zealand, Norway, Poland, Portugal, the Slovak Republic, Spain, Sweden, Switzerland, Turkey, the United Kingdom and the United States. The Commission of the European Communities takes part in the work of the OECD.

OECD Publishing disseminates widely the results of the Organisation's statistics gathering and research on economic, social and environmental issues, as well as the conventions, guidelines and standards agreed by its members.

This work is published under the responsibility of the OECD Committee for Agriculture.

Also available in French under the title:
Les coûts de mise en œuvre des politiques agricoles

Foreword

This book contains a comprehensive study of implementation costs of agricultural policies (called policy-related transaction costs). It proposes a framework to look at both public administration and economic issues, in the context of policy reform towards more decoupled and targeted policies and increasing emphasis on the multifunctionality of agriculture.

Chapter 1 of Part I identifies the different types of policy-related transaction costs that occur when implementing agricultural policies. It reviews the different steps in designing, implementing and monitoring policy mechanisms, identifies the implementation costs of the different steps, and suggests ways to reduce them, without compromising the achievement of the policy's aim. It discusses measurement issues and contains an overview of recent estimates of transaction costs, including those from three case studies presented in Part II. Chapter 2 of Part I provides a framework to include policy-related transaction costs in policy choice in order to determine the best option. It discusses trade-offs between the precision of targeting, the degree of decoupling and administration costs, using hypothetical and illustrative examples. The summary and conclusions of the whole report are presented in Chapter 3.

The OECD's Working Party on Agricultural Policies and Markets approved the publication of the report in October 2006.

Acknowledgements. Catherine Moreddu is the author of Part I and the case study on PROCAMPO in Mexico (Part II, Chapter 4). The case study on direct payments in Switzerland (Part II, Chapter 5) was prepared by Simon Buchli and Christian Flury from Flury and Giuliani GMBH, Switzerland. The case study on land conservation payments in the United States (Part II, Chapter 6) was prepared by Ralph Heimlich, Agricultural Conservation Economics, Laurel, Maryland, United States. Several colleagues have provided useful guidance and comments, in particular Carmel Cahill, Stefan Tangermann and Frank Van Tongeren.

Table of Contents

Part I
Main Report

Part II
Case Studies

List of boxes

List of tables

List of figures

THE IMPLEMENTATION COSTS OF AGRICULTURAL POLICIES – ISBN 978-92-64-03091-6 – © OECD 2007

PART I

Main Report

ISBN 978-92-64-03091-6
The Implementation Costs of Agricultural Policies
© OECD 2007

Executive Summary

There are many different kinds of costs associated with the implementation of any policy. Administrative costs, for the purposes of this study referred to as policy-related transaction costs (PRTCs), are one cost element that has been receiving attention. Concerns about PRTCs have been raised in the specific context of multifunctionality and more generally with respect to agricultural policy reform, in particular the move from market price support measures towards more decoupled and targeted policies. This study considers two main issues. The first is a public administration issue, which relates to the need to identify and track PRTCs with a view to controlling costs and making better use of public funds. The second is an economic issue, which relates to the role of PRTCs in determining the most efficient option for achieving a given policy objective.

PRTCs occur at all stages of policy implementation, from policy design and enactment to final evaluation, through interactions between and within government agencies, private organisations and programme participants. Implementation *per se* includes the delivery of payments and monitoring of eligibility and compliance, as well as the associated checks and controls.

PRTCs are necessarily incurred in the pursuit of policy objectives and are not "wasteful" *per se*, but everything else being equal, notably expected and unexpected outcomes, it will be beneficial to try to reduce them, both in order to make better use of public funds, and to minimise one of the components of the overall economic costs of a given programme. In order to reduce PRTCs while maintaining programme benefits, it is important to identify the factors that determine them. These factors relate to the characteristics of the policy, including the precision and clarity of its objectives and the nature of compliance. For a given policy, the administrative structure and the regulatory environment in place, structural factors such as the number, size and diversity of farms, and access to information and co-ordination will also be important.

PRTCs can be reduced by sharing experiences across agencies, regions or countries, exploiting already existing administrative networks, integration of government and private information systems, reducing the number of agencies, and use of information technologies. Properly measuring and monitoring PRTCs will make it easier to control them. PRTCs for a given policy can decrease over time as experience grows and initial costs are amortised.

While the importance of PRTCs in policy choice is recognised, they are rarely, if ever, taken into account in practice. The failure to take them into account is particularly noticeable in cases where big shifts in policy focus have occurred, *e.g.* from market price support to direct payments. Ideally preparations for the introduction of a new policy initiative, for example in the context of policy reform, should include a full fledged cost-benefit analysis, of which PRTCs would be a component. Transfers generated by the policy should also be considered, as how much society is prepared to pay to obtain desired outcomes is an important component of policy choice. This study found rather few attempts to estimate PRTCs. Moreover, when estimates are made, they are mostly *ex post*, with varying degrees of reliability. In order to obtain more consistent and reliable estimates for use in policy comparison, systematic and accurate procedures are needed to measure PRTCs and evaluate policies.

A full comparison of costs and benefits of different policy options needs to relate the economic value of what the policy achieves to its resource costs, including PRTCs and side-effects, as well as the transfers it may generate, both intended and unintended. This is not attempted here. In the absence of real life examples, a schematic comparative analysis is presented to illustrate the trade-offs. Stylised comparisons are carried out, for a range of hypothetical policy options, assumed to all have the same results in terms of the objective pursued. This assumption is made to simplify and to allow the analysis to focus only on the comparison of resource costs (including deadweight losses due to coupled policies, possible additional costs of de-linkage due to decoupling in the context of market failures, and PRTCs) and transfers among different policy options. Plausible assumptions regarding certain parameters and the value of unit PRTCs are made, drawing on the literature and the case studies (for example, median values from the literature reviewed in Chapter 1 are used for PRTCs). The comparisons are purely illustrative and do not represent any specific, real-life situation. This illustrative analysis indicates that, all other things being equal, the choice of policy instrument will depend on the trade-off between the targeting ratio (i.e. the share of the total transfer that is actually needed to achieve the objective) and the PRTCs. All the hypothetical examples developed for this study show that the reduction in unintended transfers as a result of targeting is one of the crucial parameters in policy choice.

Although the PRTCs of targeted payments can be higher as a percentage of transfers than those of untargeted measures, total PRTCs are not necessarily higher and in many cases, the total costs of achieving a desired policy outcome could be lower for well-targeted and well-coordinated measures. The hypothetical examples developed in this study indicate that targeted policies, whether decoupled or not, are the least-cost options under a wide range of assumptions about key parameter values, especially when the targeting ratio is low. In the case of income policies, the inclusion of income transfer efficiency in the comparison reinforces the benefits of decoupling and targeting as leakages are generally smaller (as the transfers are smaller). In the case of policies that aim to correct market failures, when jointness exists, trade-offs between gains from decoupling and the possible additional costs of de-linkage, (i.e the extra cost of producing a non-commodity output separately from commodity production – to be added to transfers to producers needed to produce it jointly) also need to be considered. This also means that the trade-off includes the transfers to producers needed for joint production of the public good on one side, and the total cost of separate production of the public good on the other side. However, the full diversity and complexity of situations in OECD countries is probably not covered, and uncertainties remain on the actual value of parameters. One could envisage cases, where implementing a targeted policy measure would not reduce the total cost of pursuing a policy objective and where the targeted option does not have the lowest cost because of high PRTCs and/or a high targeting ratio. This would presumably be the case when a policy explicitly seeks to apply a common rate of support to almost all the population, or to almost all land, and where there is no or little (negative) unintended impact, domestically or internationally. There are also cases where the total cost of pursuing a policy objective is not lower with decoupled measures than with coupled measures due to high PRTCs and/ or high additional costs of de-linkage. Finally, some governments might consider it appropriate to give different weights to welfare components and transfer components to reflect equity, feasibility and other social concerns, thus affecting policy choice.

Many issues still need to be further explored in terms of policy comparison, including the time dimension in policy implementation, the impact of institutional settings, and the other components of costs and benefits of policies. For the sake of simplification, the approach presented here compares policies that are assumed to generate the same desired outcome. In reality, the quality of the result may differ for different policy instruments and there may be other unexpected impacts, both positive and negative, that vary with the alternative policy instruments. Generally speaking, the information and data that would be needed to make these more complex comparisons are not available.

ISBN 978-92-64-03091-6
The Implementation Costs of Agricultural Policies
© OECD 2007

Introduction

In the OECD, the issue of PRTCs was raised in the context of work on the multifunctionality of agriculture. It was recognised in the analytical framework developed for the OECD's conceptual work on multifunctionality (OECD, 2001a) that when the most efficient intervention to address market failures associated with multifunctionality is sought, "transaction costs, including administrative costs associated with various options, should also be taken into account". This issue was explored at the Workshop on "Multifunctionality: Applying the OECD Analytical Framework – Guiding Policy Design" that took place on 2-3 July 2001, based on a framework paper prepared by the Secretariat (OECD, 2001b) and on two consultants' papers (Challen, 2001; Vatn, 2001). The results of the discussion that took place are incorporated and further developed in a report on the policy implications of OECD work on multifunctionality (OECD, 2003a). Another OECD report *Multifunctionality in agriculture: What role for private initiatives?* looks at non policy-related transaction costs in the context of public good provision (OECD, 2005b).

Although PRTCs attracted renewed attention in the context of multifunctionality, concerns first emerged in the more general discussion of a move from market price support measures towards direct payments in the context of agricultural policy reform. They grew further with the development of numerous, complex, environmental policy measures, targeted to specific objectives. They are now more generally expressed in terms of the costs and benefits of moving towards more decoupled and targeted payments. It is widely acknowledged that, for a given result, concerns with cost-effectiveness and efficiency of policies require all costs, including administrative costs, and benefits to be considered in determining the best option.

This study examines PRTC issues in the **broad context of agricultural policy reform**. The objectives for policy reform shared by OECD ministers include a progressive reduction of support and protection and a move towards those forms of support that are less production and trade distorting in order to let the agricultural sector respond more to market signals. OECD Ministers stressed the need for such policy reform in 1987. In 1998 OECD Agricultural Ministers agreed on a set of principles for agricultural policy reform and a set of operational criteria that should apply in designing and implementing policy measures (OECD, 1998). In particular, OECD Ministers agreed that policy measures should be transparent, "targeted to specific outcomes" and tailored, i.e. "providing transfers no greater than necessary to achieve clearly identified outcomes" (OECD, 1998). These criteria require well defined objectives that allow the elements to receive support (population, area or outcomes) to be clearly identified and the level of support required to be specific to the objectives. Ministers also agreed that policies should be flexible and equitable.

In the last decade, policy reform has focussed on changing the way in which support is provided to producers, with a notable shift away from production-linked measures that create high distortions on production and trade, towards measures that are more decoupled from current production. There has been, however, only a very modest move to policies targeted to clearly defined objectives and beneficiaries (OECD, 2005c).

Two main issues relating to PRTCs are explored in this study. The first relates to public administration and the second to the role of PRTCs in policy choice. The underlying questions are:

- To what extent can PRTCs of a given set of policy measures be reduced without compromising outcomes? How?
- Are PRTCs likely to modify policy choices? In other words, for a given result, considering all costs and benefits, can they be large enough to offset the economic benefits and transfer savings of targeted, decoupled policies?

Chapter 1 discusses public administration issues. The characteristics of the policies, the administration and regulatory environment, farm structures and information systems that affect the size of PRTCs are identified and ways of reducing PRTCs are considered. Chapter 2 discusses economic issues relating to the role of PRTCs in defining the most cost-effective and efficient option for achieving policy objectives. A simplified method for comparing policy options such as market price support and different categories of payments that are either fully coupled or fully decoupled and/or untargeted or precisely targeted, but have the same desired outcomes (results), is developed and illustrative examples are presented. Chapter 3 provides a summary and conclusions, which highlights the main issues and provides makes recommendations on the role of PRTCs in economic analysis of different policy options.

ISBN 978-92-64-03091-6
The Implementation Costs of Agricultural Policies
© OECD 2007

PART I

Chapter 1

Policy-related Transaction Costs of Agricultural Policies

1.1. Background

As explained in the introduction, this chapter examines public administration issues relating to PRTCs in the broad context of agricultural policy reform involving a move towards more decoupled and targeted policy options.

The analysis focuses on administrative costs that differ significantly from one policy option to another and therefore may affect policy choice. Setting objectives and designing solutions result in fixed costs that are common to all options whatever the decision. Similarly, the *ex ante* and *ex post* evaluation of interventions with regard to their objectives should be performed in all cases, whatever the policy chosen. In practice, governments rarely do it and past policies remain in place, without being evaluated, although underlying objectives have changed. For the sake of simplification, this study focuses mainly on the **costs of implementing and monitoring policies** – the elements most likely to differ significantly according to the policy instrument chosen and the size of the programme, and the elements found most often in the literature.

Similarly, in the context of policy reform that generally involves moving away from broad price and output support type measures to more decoupled and targeted measures, it is changes in PRTCs (and in the other relevant costs) as alternative policy options are considered that is most relevant. Different institutions and mechanisms for designing, implementing, monitoring and evaluating policies are already in place in OECD countries. To the extent possible, set-up costs are distinguished from running costs and the emphasis is given to the latter, which are more relevant to assess the longer-term cost-efficiency of a policy. Set-up costs, which can be large, can be spread over the duration of the programme.

Section 1.2 first defines PRTCs, with a view to identifying the most relevant components and determinants. It reviews the different steps in designing, implementing and monitoring policy mechanisms, and tries to identify their associated costs. Available literature on the transaction costs of policy measures is briefly reviewed in Section 1.3, and the main findings that are relevant for this study are reported. Information from the case studies contained in Part II is also included. A range of estimations of PRTCs from the literature and case studies is presented in Annex I.1. Section 1.4 discusses ways of measuring PRTCs and outlines information requirements and measurement problems. Having identified the main factors affecting PRTCs, Section 1.5 then suggests ways to reduce them, without compromising outcomes. When considering how to reduce PRTCs the context is always one in which the desired outcomes are achieved. In other words, cost savings should not be at the expense of the achievement of a programme's aims.

1.2. Definition and characteristics

From market to policy-related transaction costs

The concept of transaction costs in economic analysis was introduced in 1937 by Coase, who recognised that information on transactions is often incomplete, causing friction between suppliers and buyers. It has been broadly defined in Challen (2000) as:

"the costs of arranging a contract ex ante and monitoring and enforcing it ex post (Matthews, 1986)"; "the costs of running the economic system (Arrow, 1969)"; and "the economic equivalent of friction in physical systems (Williamson, 1985)".

Broadly defined, transaction costs could include all costs associated with any allocative decision regardless of whether the decision is made in a market or by a government (OECD, 2001b).[1] The concept has been developed in comparative economic analyses as a factor explaining the performance of various alternative institutional forms and as a parameter of institutional change. Following Furubotn and Richter (1998) and as reported in OECD (2003a), they can be classified into three categories:

- **Institutional (or political) transaction costs,** which are the costs of setting-up, maintaining and changing the institutional framework of a policy, and the costs of implementing policies.

- **Managerial transaction costs,** which are the costs of setting-up, maintaining, changing or running organisations.

- **Market transaction costs,** which are search and information costs, bargaining and decision costs, policing and enforcement costs incurred in market transactions.

This study focuses on policy-related transaction costs, *i.e.* the costs associated with the implementation of policies, and which belong to the first category of "institutional transaction costs" (Box 1.1). By definition, any policy involves some level of government intervention and therefore some level of transaction costs. This is the case for border measures such as simple tariffs, tariff quotas or export support, for domestic price support measures such as intervention purchases or production quotas, and for various types of budgetary payments and regulations. The study therefore covers all transactions between farmers and the government, whatever the implementation mechanism (payments based on eligibility criteria, contracts, regulations, taxes, etc.) or the level of government, with particular attention to budgetary payments, which have often been criticised on the basis that transaction costs are high.[2]

Box 1.1. **Terminology**

The concept of transaction costs has been developed in the economics literature relative to private transactions as a factor explaining organisational structures (*e.g.* vertical integration *versus* contracting), market failures (*e.g.* in the case of asymmetric information or a lack of property rights) and institutional choices (*e.g.* promotion of clubs). The concept, and therefore the term, has been brought in the policy debate as the cost of transactions in which the government is a party, but policy-related transaction costs are in fact the administrative costs associated with policy implementation. The term "transaction cost" has the merit of reminding us of the theoretical origin of the concept but it is sometimes confusing and to be more precise, the terms "administrative or implementation cost" are often used. In this report, for sake of continuity, we will, however, continue to use the term policy-related transaction costs, abbreviated as PRTCs.

PRTCs include the costs incurred by governments in gathering information, planning and designing policies, collecting revenue, and implementing, monitoring and checking the outcome of policies. Cost incurred by farmers when transacting with the government, to obtain information on policies and claim benefits, are also listed under this heading. It is important to include farmers' costs, to the extent possible, because they affect compliance and, in the case of voluntary programmes, participation rates. The focus is on individual costs for filling forms and talking to extension officers, rather than on collective, lobbying or organisational costs. Some of the implementation activities can be carried out by private organisations such as consulting companies, banks, insurance companies, co-operatives, farmers' organisations, Non-Governmental Organisations, certification/control associations. These activities can be partly or totally paid for by the government or be passed back to customers (i.e. farmers). To the extent possible, these costs are also considered. The study covers cases when governments use private legal instruments such as contracts. The case of quasi-markets (e.g. for emission permits) is reported in the literature review but not further developed. These issues are discussed in OECD (2005b).

The generation of PRTCs in the policy process

PRTCs are all costs arising from interactions between and within government agencies, private organisations and programme participants at all stages of policy implementation. The first thing to do is therefore to identify the specific types and stages of transactions and partners in order to characterise associated costs. What follows describes mainly the situation with budgetary payments. It also applies to market price support through tariffs only. However, in many countries, market price support is maintained through more complex border measures such as tariff quotas or export support, and domestic measures such as administered prices, intervention purchases and storage, which may also involve significant PRTCs.

The main categories of PRTCs occur first at the government level when exploring and promoting the best policy choice (research, information, communication) and when deciding policy parameters. Policies then need to be enacted and for that, efforts are sometimes made to build a consensus between the government, farmers' organisations and civil society. Implementation per se comprises the distribution of payments (or more generally support, or the enforcement of regulations) and monitoring that conditions are met. Distribution requires the identification of beneficiaries, the processing of applications, and the provision of payments. Depending on the type of policy, the processing of applications may imply further fine-tuning of specific conditions or the setting-up of individual or collective projects, which could involve specific tools such as formal contracting and technical assistance. Both eligibility and compliance, if required, need to be monitored. Enforcement and litigation actions might follow if required compliance is not met. Overall evaluation takes place at the end of the process or at regular intervals.

Figure 1.1 broadly describes transactions that can occur in the provision of budgetary payments. It classifies the different sub-categories of PRTCs depending on where they occur and who bears them in the first place. The costs incurred by farmers, farmers' organisations and non-government organisations can, however, be totally or partially refunded by the government. The diagram applies to any type of institutional structure as the various transactions can occur within or between agencies. The whole range of possible activities in policy implementation presented here does not necessarily occur. The existence of detailed sub-categories of costs and their relative importance will depend on the type of policy.[3]

Table 1.1 identifies the main PRTCs for broad types of policies. Initial setting-up costs and final evaluation costs should occur for all policies but differ depending on the number and complexity of objectives, actions required and institutions involved. For the sake of

Figure 1.1. **Sub-categories of policy-related transaction costs for the provision of budgetary payments**

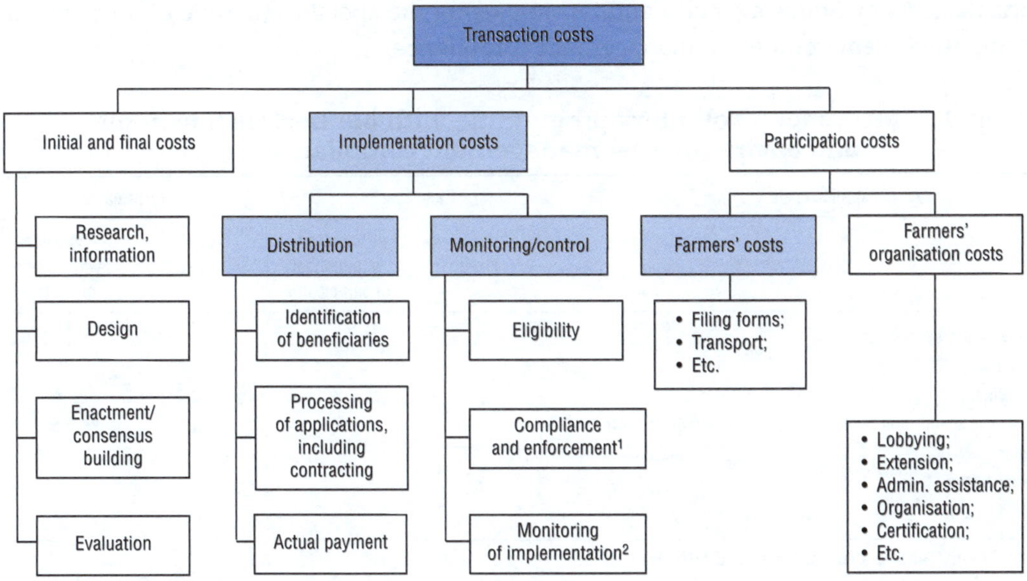

Shaded boxes contain PRTCs that are expected to be significantly different across policies and are therefore the focus of this analysis.
1. Detection of non-compliance and enforcement of compliance, litigation.
2. Includes management and organisation costs.
Source: OECD Secretariat adapted from Mann (2000).

simplification, the focus of the study is on implementation costs (shaded in the diagram). Delivery and monitoring are required for all direct payments. While output payments are received by all producers of the commodity, other payments require the processing of applications that are used to establish eligibility (declaration of area planted or animal

Table 1.1. **PRTCs of different types of policies**

PRTCs	Price support (tariff only)[1]	Untargeted, coupled payment (*e.g.* output payment)	Targeted, coupled payment[2] (*e.g.* limited output payment)	Untargeted, decoupled payment (*e.g.* pay. based on historical entitlement)	Targeted, decoupled payment (*e.g.* targeted income payment)
Research, information, design, consensus building	X	X	X	X	X
Distribution					
– Identification of beneficiaries			X		X
– Processing of applications (eligibility)			X	X	X
– Payment provision		X	X	X	X
Monitoring/control					
– Eligibility		X	X	X	X
– Compliance		X	X	X	X
Enforcement		X	X	X	X
Evaluation	X	X	X	X	X
Farmers' costs		X	X	X	X

1. In many countries, however, market price support is maintained through more complex market price interventions such as production or import quotas, export support, administered prices, intervention purchases or storage, which involve significant PRTCs. The case of such complex measures is not developed here.
2. Output payments that are specific to a less-favoured region, for example, could fall into this category. Such payments are rarely implemented.
Source: Adapted from Table 5 in OECD (2003a).

numbers). An additional selection is applied in the case of targeted payments (to specific less-favoured areas, environmental or income conditions). Table 1.2 identifies fixed and variable costs by States agencies and farmers using the specific example of a per hectare payment with environmental management compliance.

Table 1.2. **PRTCs for a voluntary programme with per hectare payments and environmental management compliance**

Administration costs	State agency		Participants	
	Fixed	Variable with number of participants	Fixed	Variable with number of hectares
Policy design (area and prescriptions)	X			
Agreement-negotiation				
– Promotion to farmers	X	X	X	
– Negotiation between agency and farmer on management requirements		X	X	X
Payment provision		X	X	X
Monitoring and evaluation	X			
Enforcement of compliance		X	X	X

Source: Adapted from Table 4.1 in Falconer and Whitby (1999a).

The actors and payers in the policy process

All organisations and individuals participating in government policy incur PRTCs. Both public and private costs need to be considered. Various government agencies are involved in policy implementation with interactions that are both vertical (between national, regional and local levels of governments) and horizontal (between networks or regions). Moreover, some of the tasks can be delegated to private institutions. Consulting companies can contribute to the ex-ante and ex post evaluation of policies. Banks or insurance companies are involved in the distribution and monitoring of support. Banks, for example, deliver loans with interest concessions. In many countries, the network and client files of insurance agencies are used to deliver insurance schemes (e.g. Spain, the United States). Extension and advisory activities can be carried out by various public, private or professional organisations. Farmers' organisations often help farmers to apply for programmes. Alternatively, consulting companies or environmental co-operatives can play this role (example of the Netherlands, see Polman, 2002). Farmers' organisations are also active in lobbying. In some countries like France, a large share of implementation activities are carried out by mixed organisations that involve both public services and farmers' organisation such as commodity boards and the CNASEA (Centre national pour l'aménagement des structures des exploitations agricoles). Farmers' organisations sometimes play an important role in coordination and certification activities. In some cases, non-governmental organisations help with the implementation of agri-environmental actions at the local level.

At the individual level, farmers interact with local government agencies, but also with their own professional organisations, mainly when filing applications and setting-up projects. They are also partners in co-ordination activities together with farmers' organisations. In the case of voluntary schemes, the PRTCs incurred by farmers are important as they can limit participation and therefore affect outcomes.

PRTCs are not always paid by those who incur them. Implementation costs by private companies are paid by customers and/or by the government, either because the government directly subsidises part or all of the administrative costs as in the case of insurance companies, or because those administrative costs are deducted from the subsidy. Similarly, farmers' organisations can receive government subsidies for their implementation activities or can be paid directly by customers (farmers). In Korea, for example, co-operatives received government payments for the costs they incurred in distributing subsidised inputs and operating intervention purchases and storage. In the case of agri-environmental or socio-structural policies, farmers' cost can be significant but they are often paid for by the programme. The next section discusses, in qualitative terms, how the distribution of costs among actors may affect PRTCs.

The different types of PRTCs

PRTCs comprise labour costs and operational expenditures.[4] Staff spend time on communication, filling forms, travelling, negotiating, setting-up projects, sending checks, controlling compliance (when the policy requires compliance), etc. Some of these tasks such as filling out forms or setting-up projects also require farmers' time. Wages and salaries for the different categories of staff, and opportunity costs in the case of farmers (often assumed to be equal to farm income[5]) are used to value unit labour costs. Operational expenditures pay for equipment such as office space, furniture, computer, printer or paper, and for services such as mailing and banking.

Irrespective of where they arise, PRTCs share various characteristics. There are fixed costs, which do not depend on the number of participants or the size of the resources involved (number of hectares or animals). Consequently, they are not affected by the size of the transfer. Variable costs, on the contrary, increase with the size of the programme. Typically, government-fixed costs occur at the early stage of the process, when the programme is being designed and promoted, and are also important in the final evaluation costs. But not all set-up and evaluation costs are fixed. Implementation costs are usually variable for both government agencies and farmers, although farmers can also incur fixed costs, for example when the programme requires the establishment of a project and the gathering of prior information (diagnosis). The distinction between fixed and variable cost is important as it affects the ability of a programme to channel large transfers.

When measuring PRTCs, the time dimension is important. Set-up costs are larger than running costs so it is important to take the implementation period of the policy examined into account when comparing policies. Ideally, set-up costs should be isolated and allocated over the duration of the policy. As experience builds up from one programme to its successors, one could also consider allocating set-up costs to a longer period than the duration of a specific programme but that would be difficult to do. The time dimension is important and would point towards the setting of longer-term policies or at least of successive policies building on each others experience.

Another consideration that is relevant when comparing policies is whether some actions generating PRTCs will not just bring about a better result, but may also reduce PRTCs in the future. The net effect would thus be positive. This can be the case of set-up costs such as technical assistance and co-ordination costs, which can be considered as investments as they help agencies and farmers improve management and should therefore result in economic benefits. Costs of gathering information, acquiring knowledge and using modern technology bring benefits that spread beyond the scope of the policy.

Finally, it would be interesting to distinguish the costs that stem from the design of the policy or the need to strengthen enforcement, and which can be justified by improved outcomes, from the costs that depend on the efficiency of delivery institutions (linked to their structure, experience, budgetary procedures, etc.). These latter costs should be minimised for any given outcome.

1.3. Review of the literature

Main issues

The issue of PRTCs is not specific to agriculture as illustrated by Box 1.2 which provides some examples of how the issue of PRTCs is considered in other sectors. This section, however, focuses on recent studies on the transaction costs of implementing agricultural policies, in particular those where quantitative estimates are provided. In the literature examined, issues concerning PRTCs are raised in the following terms:

- How to improve the design and implementation of agri-environmental and socio-structural policies in order to better achieve their objectives?

- In the context of discussion on multifunctionality, what is the trade-off between precision[6] and transaction costs?

- From an institutional point of view, which delivery systems are more cost-efficient? And finally

- How to design programmes that minimise costs for a given result, in particular through ensuring compliance and reducing cheating?

Approaches and results

The objectives, approaches and main findings of the studies reviewed are summarised here and in tabular form (Table I.1.1) in Annex I.1, which also contains estimations of PRTCs emerging from the studies reviewed (Tables I.1.1 to I.1.12) and material presented at the Workshop on Policy-Related Transaction Costs (Tables I.1.13 to I.1.19). Subsequent sections of this chapter, in particular Sections 1.4 and 1.5, are also based on information from the literature review and the workshop.

Area payments in EU member states

In the European Union (EU), the issue of administration costs of policies arose in the context of both the introduction of area payments to compensate for the reduction of intervention prices in 1992, and the evaluation of agri-environmental measures. A comparative report on the administration of arable area payments in the Netherlands, Sweden and England was published by national Supreme Audit Institutions in 2000 (SAI, 2000). Covering the period 1996-97, it examines delivery systems in five regions of the Netherlands, nine regions of England and 21 to 24 regions of Sweden, estimates PRTCs and formulates recommendations to reduce them, based on cross-region comparison within a country. The primary concern was to reduce administration costs, which are borne by member states and not reimbursed by the EU, by improving management performance of regional offices. Important regional differences were found within each country. They were partly explained by the number of claims and the share of simplified claims (below a certain number of hectares) in the region, but also by management factors, such as the level of staff training and experience, which were thought to merit further attention. In all three countries, limitations in the data available for costs and staff time were noted. National averages, which vary between EUR 168 and 381 per application, are reported in Table I.1.2.

Box 1.2. **Policy-related transaction costs in other sectors**

In what context is the issue raised?

- Administration of other sectoral or economy-wide policies, such associal, fiscal and regional policies. Regarding social policies (health, pensions, unemployment benefits), the issue is raised in terms of efficiency in targeting *versus* administration costs (Coady, 2000). Reducing fraud is of great concern. A major issue relative to social policies is also the marginal cost of raising taxes (Annex I.2). In the context of evaluating a proposed pension system of private accounts, the United States Congressional Budget Office estimated the administration costs of four pension systems (CBO, 2004). In general, given the large amounts of transfers, administration costs of social policies are relatively low (*e.g.* around 1% of unemployment benefits) in OECD countries. Governments are also concerned with the administration costs of raising taxes as explained in Annex I.2). Evaluations of regional policies (as is the case with agricultural, socio-structural and agri-environmental policies) often raise concerns about PRTCs relative to outcomes.

- Adoption of regulations: PRTCs linked to the adoption of a regulation (compliance and control costs), aimed at reducing pollution for example, are considered in other sectors as in agriculture. The main concern is to reduce control costs for a given outcome.

- Private companies implementing policies: Banks and insurance companies are involved in the implementation of policies. Concerns about their PRTCs, as in agriculture, are to minimise rent-seeking and moral hazard. Whether private management of policies is more efficient than public management has long been debated (see for example Myers, 1992).

- Investments and projects: Their implementation costs are usually carefully examined, with a view to maximise outcomes for a given total cost. Cost-benefit analyses are usually carried out.

- Foreign economic assistance (aid to developing countries, investments in foreign projects). Here too, the concern is to minimise PRTCs relative to outcomes (UNDP, 2000).

- Private charities: Because of growing concerns from donors that administration costs account for a large share of receipts, transparency has increased. The purpose here is to increase "transfer efficiency".

What can we learn? To what extent can we derive lessons for agricultural policy?

- Concerns about the design and implementation of social and regional policies are similar to those in agriculture. Lessons can be drawn in terms of targeting, as the degree of targeting (to income for example) is already greater than in agriculture.

- Similarly, there is much more experience with enforcement of regulations in other sectors and it applies on a larger scale.

- Experience with insurance and investment management is not specific to a sector.

- Foreign assistance often adopts a project approach, in particular with regard to investments. Outcomes of projects are thus the unit of reference rather than the level of payment. If project approaches were to develop in agriculture, such experience would be useful. There is also a concern to reduce PRTCs for both donors and recipients.

- The experience from charities shows that transparency is important.

General lessons

- Clear targets (well-defined and measurable objectives) are crucial in reducing PRTCs and improving outcomes.

- Involvement of many actors at all levels and wide acceptance facilitates implementation and, in particular, compliance and control (shared goals).

> **Box 1.2. Policy-related transaction costs in other sectors** *(cont.)*
>
> ● Sharing of experiences, capacity building and access to information are important elements contributing to a reduction of PRTCs.
>
> ● Using existing networks and avoiding duplication is widely advised although not always applied (in the delivering of foreign assistance for example). The importance of good infrastructure is stressed.
>
> ● Appropriate levels of centralisation/decentralisation are important.
>
> ● There are contradicting opinions about the optimal size of projects and whether a high number of participants reduces or increases PRTCs. From experiences in foreign assistance and rural development policies, some think that spreading money thinly over a large number of projects is less efficient and more costly to manage than fewer targeted large projects (UNDP, 2000). Others find small projects that can be understood at the local level easier and less costly to manage (IEEP, 2001).
>
> ● The positive role of new technologies in PRTC reduction is widely acknowledged.

Comparing agri-environmental schemes in EU member states

As part of an EU research project (STEWPOL),[7] researchers in eight EU member countries have estimated the administration costs of Countryside Stewardship Policies. The project report provides estimates of administration costs for 40 agri-environmental measures implemented under Regulation 2078/92 and Regulation 746/96. It covers a wide variety of measures, from French area payments (*prime à l'herbe*) to British customised management agreements or administratively-run (quasi) markets for environmental goods, through subsidies for organic farming (see Table I.1.1 for a list of the specific measures considered). The delivery systems, and therefore the approaches to estimate costs, vary by country and measure. It is, therefore, a rich source of information in terms of estimation methods. The summary report on administration costs (Chapter 4 in Falconer and Whitby, 1999a) also provides useful considerations about the nature and effects of transaction costs. The work starts from the recognition that these costs are not systematically a problem but lack visibility. They need to be acknowledged and set against the achievements of a given policy to "ultimately improve the value for money of public expenditures in the agri-environmental sphere". (Falconer and Whitby, 1999a.) The authors also acknowledge the importance of administration costs incurred by farmers. Falconer and Whitby (1999a) estimate the PRTCs of agri-environmental schemes at around 20-30% of total compensation payments on average, with wide variations between measures (Tables I.1.4 and I.1.5). PRTCs stem largely from factors such as the heterogeneity of producers and the asymmetry of information between participants and the government. High PRTCs are mainly found at the local level. Although evidence is scattered, PRTCs seem to be larger for agri-environmental schemes than for agricultural commodity regimes such as area and headage payments (Table I.1.3). Falconer and Whitby (1999a) emphasize, however, that PRTCs cannot be considered in absolute terms, but should be related to the outcome in terms of the policy objectives and improvements in overall social welfare (see Chapter 2 for a discussion on comparison issues).

Following from that EU report, some participant researchers further explored some issues. Falconer and Whitby (1999b) and Falconer *et al.* (2001), in particular, identify the different categories of transaction costs incurred in the implementation of voluntary schemes based on management agreements with compensation for the adoption of different management practices and the factors that affect those costs, in particular the degree of heterogeneity of farmers and agri-environmental goods, and the type of contract. They then use panel data from Environmentally Sensitive Areas (ESA) programmes (containing administration costs estimated by the National Audit Office) to test empirically the determinants of those administration costs. The number of agreements, the area under agreement and the number of years since designation are significant explanatory variables of PRTCs. The econometric analysis suggests that the incremental and cumulative numbers of agreements are important determinants of administration costs for ESA schemes. There also appears to be some scope for economies of scale related to scheme participation.

Drawing on information collected as part of the STEWPOL project, Falconer (2000) considers the impact of farmers' transaction costs on participation in voluntary agri-environmental schemes based on management agreements in EU member countries and suggests that a more integrated approach to agri-environmental schemes would contribute to a reduction in PRTCs through saving overhead costs and clarifying objectives. The long-term value of networks and capacity building for agri-environmental management should be recognised too.

Falconer and Saunders (2002) consider the trade-off between standard, site-specific management agreements between the government and farmers *versus* individually negotiated agreements, in terms of minimising total costs while optimising outcomes. Based on case-study information, they compare the PRTCs of individually negotiated and standard management agreements under the Wildlife Enhancement Scheme (WES) for Sites of Special Scientific Interest (SSSI) in the North of England. Both private and public PRTCs are considered. They find that, although standard management agreements exhibit, as expected, the lowest negotiation costs, they are more expensive with regard to both compensation and transaction costs over the whole agreement life-cycle than individually negotiated agreements. The fact that standard management agreements have been more recently implemented could explain that result, confirming thereby the significant impact of experience in implementation in reducing PRTCs.

Implementation of the Rural Development Regulation in EU member states

In the context of evaluating the Rural Development Regulation (RDR) in EU member countries, another group of studies provides general information on the implementation costs of different measures, relative to their outcomes, and identifies the factors that affect them. The studies do not, however, provide quantitative estimates of PRTCs. RDR measures in the EU (also called the second pillar of the Common Agricultural Policy, CAP) are being implemented during the period 2000-06. They encompass and go beyond the former agri-environmental measures under Regulations 2078/92 and Regulation 746/96, which were evaluated in the STEWPOL project. CNASEA (2003) looks at the implementation of both agri-environmental measures and socio-structural measures in eight EU member countries, and tries to identify the characteristics of measures that affect implementation costs. Looking at actors, institutions and attitudes towards rural development in ten EU member countries, IEEP (2001) tries to identify the factors that would foster sustainable, integrated rural development in the EU. Institutional arrangements and policy delivery mechanisms are examined, in particular. As in previous studies, the purpose is to learn from experience in implementing such programmes so as to be able to reduce implementation costs while improving results, and designing future measures that are more cost-effective and efficient. In addition, RDR measures are due to be evaluated by member countries as part as the formal EU review process before the end of the implementation period.[8]

Comparing pollution reducing measures in the United States

Various recent United States studies examine the PRTCs of policies implemented to reduce pollution levels. They all acknowledge that PRTCs need to be considered in cost-benefit analysis. The general approach is, for a given outcome (*e.g.* a given level of emission or pollution reduction), to compare the total costs of two policies. Total costs include compliance or abatement costs (foregone revenue, farmers' costs to comply plus government compensation) and transaction costs (information gathering, enactment, implementation, control, prosecution). Which alternative policy achieves the given level of pollution reduction at minimum cost? Or what is the degree of targeting that achieves the specific result at the lowest cost? Carpentier *et al.* (1998), for example, compare the costs of implementing a non-targeted *versus* a targeted pollution reduction constraint on dairies in a region of Pennsylvania. They find that, for the same reduction in nitrogen run-off, targeting based on spatial information reduces both total compliance and total transaction costs because fewer farms need to be contracted, and therefore, monitored. Thompson (1999) compares a non-tradeable emission limit permit with an effluent charge. He does not estimate all costs but only differences in compliance and (institutional) transaction costs between the two options in the case of point-source water pollution by textile mills. He finds that they do not affect the policy choice between the two measures examined and that effluent charges are still clearly preferable to non-tradable effluent limit permits when PRTCs are taken into account. He concludes that PRTCs did not modify the choice given by traditional cost-benefit analysis but that they are significant and depend on the institutional framework (notably lobbying costs). Consequently, efforts should be made to estimate them better. McCann and Easter (1999 and 2000) estimate the transaction costs of policies to reduce agricultural phosphorous pollution in the Minnesota river; and the public sector transaction costs in NRCS (National Resource Conservation Service) programmes (Table I.1.6). They find that transaction costs vary widely between policies. While an input tax is more efficient at reducing pollution than a regulation on production methods in the absence of transaction costs, its advantage is even greater when transaction costs are taken into account.

Comparing institutional delivery systems

Looking at the efficiency of institutional delivery systems, Mann (2000) proposes a methodology to estimate the transaction costs associated with the implementation and administration of public programmes and applies it to investment policies (grants and reduced tax rates) in three regions of Austria, Germany and Switzerland, with different institutional settings. His estimates are presented in Table I.1.7. Distribution costs, control costs, consulting costs, implementation costs, applicant's costs and lobbying costs are included to some degree. He also considers, from a theoretical point of view, the relationship between transaction costs, transfers and results. In another study, Mann (2001) examines inter-organisational efficiency in agricultural policy administration in 131 regions of Germany, focusing on factors that relate to the structure of the administrative system. He estimates the administration costs of all agencies involved in policy implementation and tests which factors explain PRTCs per hectare or per farm. Factors examined include farm size, contracting out of some activities, absence of a specific regional Ministry of agriculture, number of offices a farmer has to deal with in a region, existence of small regional offices matching one county, involvement of Chambers of Agriculture and existence of a district office (see PRTC estimates in Table I.1.8). According to Mann (2001), the two main factors increasing PRTCs are the multiplication of local (horizontal) agencies and the inclusion of an additional vertical layer (here a district office) that increases the number of partners and therefore transactions. Mann (2002) then adapts the concept of administrative elasticity, which links the administration costs of a programme with the monetary volume of transfers, to agricultural policies, and applies this concept to the administration of export subsidies in Germany (see PRTC estimates in Table I.1.9). The concept of administrative elasticity should be used with caution. The time element is crucial and only long-term elasticities are relevant. Administrative systems are proved to be inelastic as they do not adjust quickly to changes in the volume of tasks that is often variable in the implementation of export subsidies.

Similar issues about institutional efficiency at delivering programmes are raised by Ker (2001), who discusses the relative PRTCs of government *versus* private insurance companies in the delivery of insurance programmes in the United States and Canada. Ker argues that although, in theory, insurance companies should be more efficient at delivering insurance programmes than the government because of pre-existing channels and asymmetric information, if such administration costs are partly paid for by the government, companies could adopt rent-seeking behaviour and those benefits would not materialise. In fact, the government-delivered Canadian crop insurance programme has lower administration costs, relative to premiums, than the crop insurance programmes in the United States delivered by private companies, but the level of take-up could also explain that result (Table I.1.10). Other estimates of the PRTCs of insurance programmes are presented in Table I.1.11.

Limiting moral hazard and control costs

Another relevant stream of literature examines policy design and parameters that could limit moral hazard[9] and therefore reduce control costs (Fraser, 2004; Latacz-Lohmann and Van der Hamsvoort, 1998; Millock *et al.*, 2001). Concerns about reducing moral hazard have been widely considered, in particular, in relation to insurance programmes (Miranda, 1991; Skees, 2000; OECD, 2000b; Weaver and Kim, 2002).

Multifunctionality and policy choice

In the context of work on multifunctionnality, Vatn *et al.* (2002) discuss best policy choices, depending on the degree and type of jointness between commodity and non-commodity outputs and the transaction costs of policy implementation. They develop a theoretical model to evaluate the trade-off between transaction costs and precision. Based on interviews, they estimate transaction costs for a wide range of Norwegian policies: price support for milk, environmental tax on mineral fertilisers, environmental tax on pesticides, price support on home-refined dairy products, acreage payments, subsidies for reduced tillage, acreage support for organic farming, support for preserving cattle races, support for special landscape ventures (Table I.1.12). The coverage of PRTCs is large. All levels of government costs, importers, processors, co-operatives and farmers' costs are included. As most programmes have been in place for a while, running costs are measured and not set-up costs. For sake of comparison, it would have been useful to have disaggregated information by type of PRTCs that are collected (implementation, control, etc.). It can be discussed for example whether the collection of information is a specific PRTC as it is needed to make an informed choice and assess outcomes whatever the policy (*e.g.* micro-level data, environmental indicators). In Vatn *et al.* (2002) measures attached to commodities are found to have lower PRTCs as a percentage of transfers because their asset specificity (*i.e.* the level of specificity or targeting of the outcome) is low and transaction frequency relatively high, while highest PRTCs as a percentage of transfers are found with measures applied to non-commodity outputs that are site specific and have low frequency (Table I.1.12). The authors conclude that asset specificity is the major determinant of PRTCs. As expected, measures which involve high transfers are found to have lower PRTCs as a percentage of transfers (see Chapter 2 for a discussion on the appropriate unit for comparing policies).

Also, in the context of work on multifunctionality, OECD (2003a) discusses the information needed to make a correct comparison between various policy and non-policy options for the provision of public goods. The types of PRTCs associated with the implementation of targeted payments, output subsidies and price support through tariff are identified. It is acknowledged that transaction costs should be measured early in the sequence for policy-makers to take them into account when designing policies.

Lessons

Concerns about PRTCs are growing and lessons from policy implementation are being drawn in terms of reducing PRTCs for both farmers and governments. No examples have, however, been found in the literature of where PRTCs have been analysed along with wider economic costs of policies, as a basis for making an informed policy choice. There have been a number of attempts to estimate PRTCs in the literature but evidence is scattered and focuses on agri-environmental policies linked to the provision of non-commodity outputs and/or the reduction of negative externalities. Reviewed studies focus on the comparison of costs (and sometimes benefits) across measures and the identification of factors affecting PRTCs.

The different categories of PRTCs have been characterised clearly and, although differences in terminology are sometimes confusing, a common typology can be proposed (in Section 1.2). Every type of measure and delivery system will have a different combination and relative importance of costs.

Relevant approaches to estimate PRTCs depend on the information that is available to analysts given the time and resources at hand, but also on the nature of the programme. Various types of approaches are used in different conditions (see Section 1.4). The often indirect nature of estimations and the difficulties attached to each method, whether based on interviews, or estimated using labour costs or budgetary information and flow charts, are outlined in Section 1.4. In addition, cost coverage is sometimes partial: it only includes labour costs (time multiplied by wages, often the main component), but ignores equipment costs. If policy makers were to take PRTCs into account when making policy choices, more systematic and consistent estimations would need to be made. To that end, it would be useful to develop a methodology that would register costs as they occur.

There has been much consideration of the factors that affect PRTCs, in qualitative terms, but also in quantitative terms. The experiences reported in the studies are clearly useful and Section 1.5 draws heavily on them. Econometric estimates to explain PRTCs require a wide coverage of regions and measures. Evidence is not so clear as many factors affect PRTCs. Attempts have focussed on policy design and the characteristics of participants, but also on the institutional framework.

Three types of comparisons have been attempted. The first one consists in comparing PRTCs between regions and countries, with a view to identifying the factors that influence them and to recommending changes in the institutional delivery system in order to reduce PRTCs (SAI, 2000; Mann, 2001, 2002). The second type of comparison is between variations in design of a specific policy (agri-environmental or insurance schemes) in order to minimise delivery and control costs (*e.g.* Ker, 2001; Thompson, 1999). Finally, the third type consists in comparing PRTCs over a large set of policies (Falconer and Whitby, 1999a; Vatn *et al.*, 2002). The scarcity of attempts at comparing a wider range of policies, including broad-based ones, underlines the challenge in terms of finding a common unit and assessing the other costs and benefits, notably outcomes. Cross-country comparisons for a given type of measure have also been attempted but also raise serious issues in terms of policy comparison, such as differences in the institutional setting and efficiency of delivery systems, level of take-up, exchange rates and wages. Many issues still need to be further explored in terms of comparison. The time dimension would, in particular, require more attention, as well as the other components of costs and benefits of policies that achieve the same objective.

To complement and update estimations from the literature, three country case studies were undertaken as part of this study. The first estimates the different components of the PRTCs of PROCAMPO in Mexico and illustrates how they can be reduced (Chapter 4). The second compares the PRTCs of two types of direct payments (general payments with cross compliance and ecological payments) in two regions of Switzerland, Canton Grisons and Canton Zurich (Chapter 5). The third examines the evolution of some of the implementation costs of the Conservation Reserve Program in the United States and compares them to the costs of other land conservation programmes (Chapter 6). Estimations of PRTCs from these case studies are summarised in Table I.1.13 to I.1.17. A summary table (Table 1.3) presents the range of PRTC estimations as a percentage of payments, found in the literature and case studies, for various types of policies.

The following sections of this chapter draw on the literature review to provide a definition of PRTCs, to identify the factors affecting them, to discuss the scope for reduction and to propose methodologies to estimate them. Chapter 2 examines issues raised when comparing policies, and discusses the role of PRTCs in policy choice, the central question of this study.

Table 1.3. **Selected examples of PRTCs as a percentage of transfers for various policies in different countries (selected from Annex I.1)**

Policy	Minimum	Maximum
Export subsidies[2]	0.44	2.26
Output payments in Norway[1]	0.25	12
Area payments (EU)[3, 4]	0.8	6.8
Headage payments (EU)[4]	2.5	20
Organic payments (EU)[5]	2.4	30
Environmental payments in the EU[6]	1.1	574[12]
Area payments in Switzerland[7]	0.3	2.2
Livestock payments in Switzerland[7]	0.5	0.8
Organic payments in Switzerland[7]	0.9	1.6
Ecological compensation in Switzerland[7]	2.3	2.8
Animal welfare payments in Switzerland[7]	1.5	4
PROCAMPO payments[8]	2.9	2.9
Investments subsidies[9]	13	52
Insurance payments[10]	15	357
Land conservation programmes in the US[11]	5	41
Land use/landscape payment[1, 6]	8	110

These estimates provide a range of values for PRTCs. They were obtained using different methodologies in countries with different institutional environments and for programmes including different conditions. Data are therefore not comparable.

1. Table I.1.12.
2. Table I.1.9.
3. Table I.1.2.
4. Table I.1.3.
5. Table I.1.5.
6. Table I.1.4.
7. Table I.1.17.
8. Table I.1.13.
9. Table I.1.7.
10. Tables I.1.10 and I.1.11.
11. Tables I.1.6 and I.1.14.
12. Such high estimates of PRTCs per unit of transfers are found for very particular schemes that do not channel large transfers, either because they are very limited in scope and/or because they are newly introduced. In the latter case, high set-up costs, including capacity building, contribute to such high estimates.

Source: Annex I.1.

1.4. Measuring policy-related transaction costs

It is clear that PRTCs need to be taken into account when designing optimal policies and that transparency helps to reduce them, from which follows the importance of measuring them. In other words, the measurement cost can be considered as a PRTC, whose potential benefits in terms of reducing future PRTCs can offset the initial cost.

There is generally no systematic reporting of PRTCs. However, in recent years, researchers have attempted to estimate PRTCs, covering a wide range of measures but mainly focusing on agri-environmental schemes. Drawing on that experience, some methods are suggested below. The choice of method will depend on available information and type of transaction.

It should be acknowledged that measuring PRTCs is not easy and can be time consuming. That's why, in terms of measurement, the study suggests focusing on the elements that differ between policies. Here again, there is a trade-off between accuracy and measurement cost.

THE IMPLEMENTATION COSTS OF AGRICULTURAL POLICIES – ISBN 978-92-64-03091-6 – © OECD 2007

Information requirements

Implementation agencies

Information is needed on labour costs and operational expenditures. This is generally available at the aggregate level, from the budget of the agencies. The difficulty is to allocate these costs by measure and by type of PRTC.

Most information on the allocation of PRTCs lies within government agencies and is not easily accessible by researchers or the public, partly because it is not always available in a synthetic and user-friendly format. The methods to estimate and allocate PRTCs will depend on the institutional framework for policy implementation, on the type of policy and on available information.

To estimate the different categories of PRTCs for each policy, information is needed on:

- The different types of policy measures in a country and the implementation sequence, *i.e.* the different categories of PRTCs incurred.

- The various organisations involved in the delivery and control of the policy at all levels of government.

- Budget information on administration costs for each organisation (there are usually one or two lines per agency: personnel costs and operational costs) and the amount of transfers.

- The structure of each organisation (flow chart), *i.e.* the different administrative sections or units.

- The time spent on tasks and measures by sub-section (according to set objectives and from surveys or questionnaires on individual tasks.

- The number of staff in each category by sub-section.

- The average salary of each staff category.

Information on tasks and staff is often available from the description of the policies (purpose, implementation network) and the organisations' purpose (mission) and flow chart. Information on average wages by category of administrative staff is widely available. The tasks performed by each sub-section can sometimes also be obtained from the general description of the institution and its objectives. However, the allocation of tasks between staff within one sub-section and for one individual carrying out several tasks or dealing with several measures requires more detailed investigation from surveys/questionnaires on how time is allocated. In the absence of such detailed information, approximations can be made.

Contracted agencies such as banks and insurance companies

The government usually pays for the administration costs contracted agencies face when implementing policies. This payment is usually identified in the government budget. It can be used as an estimation of their costs. Such information should however be interpreted with care as private companies estimations cannot be double-checked and rent seeking behaviour might lead them to overestimate such costs. This has often been the case when private agencies implement government-funded programmes, such as insurance schemes (OECD, 2000b). With appropriate government procurement practices, however, the administration cost of agencies should represent the market cost for the delivery of the contracted service.

Farmers

As for government staff, PRTCs for farmers include the time they spent filling forms, travelling, checking compliance, multiplied by the opportunity cost of their time, *i.e.* farm operating income. The purchase of equipment and payments for services (stamps, advisory services) are also part of the PRTCs. They can only be estimated by farmers themselves and reported in interviews or questionnaires.

Measuring the PRTCs of existing policies

Possible *ex post* methods are explained below. They are methods found in the literature that have been used by researchers and are based on existing information. To obtain more consistent and precise estimates, it might be, however, preferable to register PRTCs as they occur, as was done for agri-environmental payments in the United Kingdom, and/or to use a common method, such as the Standard Cost Model described below.

Direct estimation

PRTCs can be estimated directly through interviews or questionnaires. Implementation agents and farmers are asked how much time they spent on each task and measure. This time is multiplied by average wages for each category of staff (or farm income in the case of farmers) to get an estimation of the value of costs. Operational expenditures (*e.g.* communication material, mailing) can also be obtained through interviews or questionnaires. Vatn *et al.* (2002) seem to use this kind of direct method. It is time consuming when carried out on an *ad hoc* basis but could be done regularly as PRTCs occur, as part of the policy implementation process, as was done in England for some agri-environmental programmes for a period (Falconer *et al.*, 2001).

Top down approach

The starting point is to list all the organisations involved in implementation, monitoring and control, at all levels of government and collect information on their **total administration costs** as appears in the organisation's budget.

The total cost of an organisation then needs to be allocated to the different tasks (*i.e.* types of transactions) and measures. This can be done using the relative shares in labour costs derived from the information mentioned above on the different tasks performed, number of staff and measures dealt with by the different administrative sub-sections.

When individual PRTCs by task and measure have been estimated for each organisation, the equivalent costs can be summed up over all the organisations. This method is used, for example, by Mann (2000, 2001 and 2002). This method was used to estimate ASERCA's cost in implementing PROCAMPO in Mexico (Chapter 4).

Bottom up approach

Information from detailed studies of a few "typical" cases is first gathered. The characteristics of the population (of contracts) are then analysed and compared to the "typical" case, allowing for generalisation. This method has been used to estimate staff and farmers' costs for the implementation of complex, long-term contracts (see for example Falconer and Saunders, 2002). It is suitable when the number of people involved is large, for example for local agents and farmers, as it helps to reduce estimation costs.

A possible standard method: The Standard Cost Model

For sake of consistency and transparency, it would be useful to develop guidelines or standard procedures on measurement methods for estimating PRTCs. Based on practical experiences, some countries have developed a standard cost model, which has been used to estimate the administrative burden that regulations put on businesses. It is summarised in Box 1.3. This type of model could be adapted to develop a methodology to estimate PRTCs in a more consistent and systematic way.

Box 1.3. The Standard Cost Model: A framework for defining and quantifying administrative burdens for business

The Standard Cost Model is a framework defining and quantifying administrative burdens for businesses. It was developed by an international working group on administrative burdens consisting of Denmark, Sweden, Norway, Belgium, the UK and the Netherlands. It contains the framework for a quantitative methodology that can be applied in all countries. It first defines the administrative burdens (AB), whose cost is to be measured. The individual activities to be carried out then need to be identified. The number of times activity i is performed (Q_i) depends on the number of businesses (B_i) and the frequency of the activity (F_i): $[Q_i = B_i * F_i]$. The cost of the single activity i (C_i) depends on the time required to perform it (T_i) and the per unit cost (P_i): $[C_i = T_i * P_i]$. The unit cost of activities carried out internally is based on employees' wages, and material and overhead costs, while the cost of activities carried out externally are based on the costs of contracting out. The administrative burden of all activities is therefore the sum of costs for individual activities: $AB = sum(i) [Q_i \ C_i]$.

Source: International Working Group on Administrative Burdens, *The Standard Cost Model*, August 2004. Further information on administrative burdens is available at *www.compliancecosts.com*. Presentation made in Session 3 of the OECD Workshop on Policy-Related Transaction Costs (OECD, 2005a) on information needs and measurement issues by Peter Ladegaard, OECD, Public Governance and Territorial Development Directorate, Regulatory Management and Reform Division.

Measurement issues

Administration costs as reported in the budget or estimated by staff labour costs do not always correspond to the PRTCs of interest for several reasons. Inflexibility in administrative structures (staffing) and imperfect predictions of participation rates (number of farmers) means that staffing levels may not be efficient at any given time. Long-term costs have to be considered. Under-staffing (to make administration costs look lower or following a policy change) can happen at times but would not be sustainable in the long run. It is difficult to distinguish set-up costs from running costs. High initial costs may be an investment that allows for later savings in costs (innovative technology for example). Answers from direct interviews and questionnaires also have their bias as people tend to overestimate their efficiency and underestimate PRTCs. Conversely, managers could overestimate PRTCs to get more resources. Direct registration of costs as they occur is likely to lower this type of bias but not eliminate it. In the case of activities contracted out to non-public agencies, compensation costs do not only reflect PRTCs incurred by those agencies but also their bargaining power in dealing with the government.

Estimation of PRTCs for new policies

When designing policies, some estimation of PRTCs, based on previous experience, is needed. The likely PRTCs of a new policy can be directly estimated from interviews with staff who will implement them, in order to take account of the specific institutional setting. However, results should be interpreted with caution as the choice of staff may influence the answer (McCann and Easter, 1999).

PRTCs from existing policies can be used to estimate the PRTCs of a new policy by comparison of the various factors that affect the different costs such as the policy characteristics, the implementation network, the expected take-up level, etc. These factors are outlined in Section 1.5. More formally, there have been attempts to explain PRTCs using such factors as explanatory variables in econometric regressions (see for example, Falconer and Whitby, 1999b; Falconer, 2000; Mann, 2001; McCann and Easter, 2000). Such equations could be used to predict the PRTCs of new policies. Optimisation models that include PRTCs can also be used. Carpentier *et al.* (1998), for example, use a linear programming model to determine optimal targeting, for a given level of PRTCs. A range of possible PRTCs can then be used (sensitivity analyses). It is therefore essential to have detailed information on existing PRTCs to be able to estimate future ones and design cost-efficient future policies.

Indicators of PRTCs

Once PRTCs have been estimated, how can they be used and interpreted by policy-makers? It is not at all appropriate to compare the total value of PRTCs in absolute terms across policies. First, they depend on the size of the policy, which can be measured in terms of transfers, outcomes, number of participants or contracts, land coverage, etc. Second, they are only one factor in a complex array of costs and benefits.

Ideally, PRTCs should be expressed in relation to the value of the change induced by the policy, *i.e.* the unit of output multiplied by the social value of the marginal unit. Output is defined in broad terms to include all costs and benefits, including the outcome with respect to the objectives set and wanted or unwanted side-effects.[10]

The multiple outcomes of a policy are difficult to value, in particular positive and negative externalities because in the absence of markets, their value cannot be directly determined.[11] In an intermediary step, for the sake of presentation and to avoid valuation problems, PRTCs could be related to quantitative indicators of desired outcomes, which would be set in policy objectives. PRTCs could for example be expressed in relation to the number of kilometres of hedges planted, or tonnes of organic milk produced, etc. In the literature, PRTCs are also related to the number of participants, in particular when filling a form or signing a contract is required. They are also expressed per hectare, in the case of land-based policies (for example, the number of hectares covered by a land management scheme or farmed organically). All these indicators could be considered as proxies for outcomes.

In order to eliminate the size effect, PRTCs are often expressed as a percentage of transfers. The assumption in doing so is that transfers are a proxy for the size of the policy or the value of social outcomes. When considering direct payment programmes, expressing PRTCs as a percentage of payments also provides a complete picture of how public funds are used. This is, however, not an appropriate unit for comparing policies as it does not generally reflect all costs and benefits. This type of partial comparison should in addition be interpreted with care so as not to give the impression that transfers themselves are the objective of the policy and that increasing them is a legitimate way to reduce relative PRTCs. It is true that PRTCs can be reduced when participation (and therefore transfers) increase, but such a comparison does not imply that the level of support should be raised. When transfers pay for farmers' implementation costs, an additional difficulty in interpretation is that PRTCs are in both the numerator and the denominator.

In summary, it is clear that PRTCs as a percentage of transfers should not, alone, determine policy choices. Given the difficulty in evaluating policy outcomes, Chapter 2 examines how PRTCs can be taken into account in policy choice.

1.5. Reducing policy-related transaction costs

All government programmes incur PRTCs. They must be incurred in order to ensure that policy objectives are met. Some can even be considered as beneficial investment (i.e. its longer-term benefits are expected to offset costs as in capacity building and technical assistance). Efforts should, however, be made to minimise them as long as outcomes are not affected, in order to make better use of public money and improve administrative efficiency. Typically, policy funding is not unrestricted, and hence the positive opportunity cost associated with using these funds should be weighed against alternative uses. Reducing PRTCs would also contribute to improving economic efficiency. In terms of public administration, distributional issues also arise as implementation costs may be borne by other parties than those who design policies. This issue becomes more relevant as targeted policies increasingly require decentralised and local implementation. In the case of voluntary schemes, keeping PRTCs for farmers to a minimum will improve the rate of adoption of the scheme.

The first step is to identify factors affecting the size of PRTCs. The second step is to envisage ways to act on these factors in order to reduce PRTCs, without compromising the outcomes of the policy.

Factors affecting the size of PRTCs

Many factors affect the total size of PRTCs, through their impact on the number of transactions, their frequency, and the cost of each transaction (unit PRTC).[12] They have been identified, both theoretically and empirically, in the literature (as mentioned in Table I.1.1), and there have been several attempts to test their explanatory power. Factors affecting the number and unit cost of policy-related transactions are outlined below as pertaining to four broad categories: the characteristics of the policy, the institutional environment, structural factors and information/coordination systems. As some factors can have opposite effects on different sub-categories of PRTCs, those specific impacts will be identified to the extent possible. It should also be kept in mind that some categories of PRTCs, such as those linked to compliance control, do not apply to all policies. They only occur in the case of payments that are conditional.

Policy characteristics

As total PRTCs increase with the number of transactions, policies requiring a high number of interactions, for example when there is a need for technical assistance, co-ordination, or heavy monitoring procedures, will therefore have higher PRTCs (other things being equal). Frequent or regular interactions affect PRTCs because they allow for systematization and therefore a reduction of unit PRTCs. Policy design can help reduce the number of transactions and increase the frequency of standard transactions (thereby reducing the unit cost), although institutional and structural factors also matter.

Experience has shown that policies with **clear, well-defined, measurable objectives** are less costly to implement and monitor because all actors understand the conditions and are more likely to support objectives that have been quantified and explained. As a result, there will be fewer interactions in communication, extension, implementation and monitoring. As part of the definition of objectives, adopting a small number of clear, measurable criteria helps to reduce control cost as compliance is easier to observe. The type of requirement influences the number of transactions and the cost of each transaction. Targeting outcomes rather than processes for example leads to lower evaluation costs. In the case of conservation schemes, structural practices (such as building terraces) have lower PRTCs than management practices because the number of transactions is lower and also because compliance is easier to observe, unit costs are lower.

Precision in the definition of objectives (i.e. targeting) also affect PRTCs. Some authors (Vatn et al., 2002; McCann and Easter, 1999, 2000) use the term **asset specificity** to refer to the qualitative aspect of the outcomes, i.e. the level of specificity or the degree of targeting (to agents and site). PRTCs are often thought to be higher if the output is heterogeneous, specific to areas, practices, etc.; conversely, they are thought to be lower if the output is homogeneous and standardized routine transactions can be performed, leading to potential economies of scale. Asset specificity often implies low frequency of transactions, potentially reinforcing these effects. All other things being equal, unit PRTCs of tailored, specific programmes can be limited when competitive bidding is used to select projects and GIS technology is used to control compliance. In fact, various factors work in opposite directions and the resulting effects need to be carefully evaluated. Targeting may increase unit information costs, although whatever the policy, information should be collected anyway to make an informed choice and to evaluate results. If the number of transactions per farmer is higher with targeted measures, the number of applications or negotiated contracts is smaller and targeted farmers can be expected to be more homogenous. As a result, total information, implementation and monitoring costs may be lower as they only apply to targeted farms. On the other hand, standardization of procedures (and therefore a lower degree of targeting) lowers unit PRTCs but leads to over-compensation of the farmers who are the most efficient at providing the goods and services being paid for. In order to lower total PRTCs and improve outcomes more efficiently, some authors recommend concentrating on fewer measures that yield large benefits or on areas where benefits are larger.

In the context of the implementation of agri-environmental or structural measures, the issue concerning the degree of targeting is often expressed in terms of comparing **bottom-up** (project) approaches with **top-down** (guichet) approaches (CNASEA, 2003). Bottom-up approaches have higher implementation costs as a higher number of transactions is required to set up a project and the unit cost increases with the complexity of the project, but they are expected to have lower control costs as participants understand and agree to the conditions set. In addition, they are probably more efficient in achieving desired outcomes and are therefore often favoured by rural development actors. Top-down approaches on the other hand often have lower implementation costs as transactions are standardized, but they have higher control costs (compliance is lower because the requirements do not reflect the specific conditions) and possibly lower efficiency (the targeting is often less precise because of information asymmetries or deficiencies).

PRTCs can be reduced in relative terms if the **number of participants** can be raised, *i.e.* if economies of scale can be made. Variable costs per transaction can thus be reduced and fixed costs can be spread over a larger volume of transactions. Similarly, PRTCs are expected to decrease with the **duration of the programme** as experience (capacity building) should lead to a reduction of variable costs and fixed costs can be spread over a longer period. Conversely, where policies are changed frequently, PRTCs will tend to be larger.

More generally, ways to **improve compliance** (for policies where compliance is required) through the design of the policy by reducing the risk of moral hazard may also reduce monitoring and control costs. Simplification of procedures and conditions will reduce implementation costs as requirements can be more easily understood by farmers without the need for expert assistance. Conversely, more stringent requirements will increase the risk of moral hazard for a given probability of detection. The alternatives are precise requirements involving a high degree of monitoring, which would improve outcomes but raise PRTCs, or more flexible implementation and control procedures and conditions that would facilitate adoption and therefore also improve overall outcomes although less precisely, with lower PRTCs. It has been observed that better results can be achieved with flexible measures when objectives are clear and shared by all actors. Suggestions for reducing moral hazard by reducing information asymmetry and therefore control costs are given in the next sub-section.

If the same survey, information gathering and monitoring systems can be used for several different measures, there will be savings. **Policy integration**, which avoids contradictory measures and defines priorities clearly, may also reduce PRTCs as well as improve results. But on the other hand, PRTCs might increase with the need to co-ordinate policies and prevent over-compensation. Programmes with a high number of measures have proved difficult to implement, in particular when a large choice of options is offered because many alternatives have to be considered. As a result, negotiation and enforcement costs are high. At all levels, **transparency** helps to reduce PRTCs because it increases information availability and reduces information asymmetry.

When examining project implementation, the **size of projects** has also been discussed in agriculture, in the same terms as indicated in Box 1.2 for other sectors. Some authors favour smaller projects because they are less complicated to implement, and they can be more flexible and innovative. Conversely, CNASEA (2003) and UNDP (2000) find that collective, larger projects are less costly to implement and yield greater benefits than a large number of small individual projects.

Administrative and regulatory environment

The existing **policy environment** is a major factor as most PRTCs, in particular set-up, information or monitoring costs, can be reduced using existing networks and human experience. Most countries reform their policies gradually and make use of existing delivery systems when implementing new policies.[13] The integration of agricultural and non-agricultural institutions, using a common delivery framework would also contribute to reducing PRTCs. Agri-environmental or structural policies could use similar channels as regional, rural or environmental policies. Income support could be delivered through the general tax or social security systems, with little additional marginal costs. A good rural infrastructure can help to reduce PRTCs as it facilitates travel, information gathering and co-ordination.

Transaction costs are sometimes defined as the costs to clarify **property rights** (Allen, 1991). Clear property rights reduce PRTCs. Unclear property rights, as in the case of non-point source pollution, lead to higher PRTCs. It follows that a policy that requires a change in property rights might encounter resistance and therefore creates high PRTCs, at least initially.[14]

The administrative environment affects both the number of transactions, which increases with the **number of agencies**, and the unit cost of transactions, depending on the efficiency of each agency.

Various agencies can successively (vertically) intervene in policy implementation from the central to the local level. The level of **centralisation/decentralisation** depends on the general administrative structure in the country. It is acknowledged that a higher degree of decentralisation allows for a better definition of policy objectives and increased trust at the local level, which, as seen above, could contribute to a reduction of PRTCs. However, the accumulation of administrative layers can increase implementation costs and sometimes leads to a duplication of tasks (mainly in terms of monitoring and control) if responsibilities have not been well defined.

The **horizontal administrative structure** of the country ("*découpage administratif*") also affects the horizontal structure of agencies covering the territory and therefore the number of local agencies involved in policy implementation. The optimal size of local agencies is when they are small enough to have some homogeneity and close contact with all actors, but large enough to allow for economies of scale. In fact, the multiplication of small agencies can greatly increase PRTCs (Mann, 2002). This is very important as the highest costs occur at the local level. On the other end, large constituencies increase travel time and expenses.

The administrative structure may influence the **distribution** of PRTCs among the levels of government and other actors. As seen above, the size of PRTCs will depend on who pays the implementation costs and who pays the transfers. If PRTCs are not paid ultimately by those incurring them and if the paying agency has no control over them, there is no incentive to reduce costs. There is even an incentive to adopt rent-seeking behaviour and overestimate administration costs. This is often the case when government delegates some tasks to non-government agencies in a non competitive environment. An institutional framework where PRTCs are not borne by those incurring them will therefore lead to higher PRTCs. Such a system could also influence the policy choice. Conversely, PRTCs that are passed onto scheme participants through certification or reporting requirements are expected to be lower.

Structural factors

The **number of beneficiaries** or agreements affects PRTCs both ways. On the one hand, fixed costs are reduced in relative terms if they can be spread over a larger number of recipients. On the other hand, for a given transfer, variable costs increase, by definition, with the number of recipients.

As seen in the policy characteristics section, the ability to carry out homogenous transactions reduces unit PRTCs. This depends on the policy design but also on structural factors such as the **homogeneity or diversity** of situations (regions/sites, agricultural systems, farm structures). Homogeneous structures will facilitate the provision of a standardised outcome, using standardised transactions, and therefore reduce PRTCs.

Additionally, as standard transactions will better fit the situation, compliance will be easier, thereby also lowering moral hazard and control costs. Structures also affect the cost of compliance, which in turn affects PRTCs. Homogeneous structures facilitate compliance to a relatively standard measure that would be cheaper to implement. In turn, easier, low cost compliance facilitates adoption and therefore requires less control.

Physical farm size also affects PRTCs. If the outcome of the policy depends on the number of hectares (or animals), fixed PRTCs can be spread over more hectares (or animals) if the farm is big, resulting in lower PRTCs per outcome. In other words, in order to achieve the same outcome (number of hectares under conservation programmes for example), PRTCs will be higher if forms have to be filled or projects and contracts have to be made with more farmers. Accordingly, small farms incur higher PRTCs for a given result and as a percentage of benefits. This, in addition to higher compliance costs, could in part explain their lower participation in voluntary environmental schemes. In recognition of the effect of physical size on PRTCs, some programmes include a simplified delivery system for smaller farms. This is the case for area payments in the EU.

Information/co-ordination systems, including experience, technology

Time has been identified as one major factor influencing PRTCs. Initial, set-up costs can be large. However, they need to be spread over the implementation period (when comparing policies). That's why the duration of the policy is important. The set-up phase can be seen as an investment. Falconer suggests that capacity building is an objective to be funded in itself. Moreover, consensus builds over time so control costs should decrease, especially for co-operative approaches. Growing administrative experience allows for fine-tuning of implementation and improvements in efficiency (including through the use of technology).

Information asymmetries and deficiencies increase PRTCs. Information is costly to get and control costs are higher (Vatn *et al.*, 2002; Fraser, 2004; Falconer and Whitby, 1999a, p. 72). The level of uncertainty (about the practices used and the outcome) increases information and control costs.

Information technology can significantly reduce routine PRTCs (Box 1.4).

Political acceptance and recognition of the overall objective has already been mentioned as a factor reducing PRTCs in the context of designing clear objectives but it is a more general issue. Control costs are lower because compliance is better but communication efforts are needed that generate PRTCs. Conversely, resistance to change increases PRTCs. Involvement of all actors at an early stage of the design, active local associations and trust between stakeholders, have been identified as factors lowering implementation PRTCs (because they improve compliance and reduce control costs), although they increase set-up costs. The sharing of experiences and good practices also reduces PRTCs. Informal factors, although difficult to grasp, can significantly affect political acceptance, and therefore the behaviour, of both farmers and delivery agents. Pre-existing communal efforts and networks facilitate the implementation of common projects.

Identification of factors affecting PRTCs naturally leads to recommendations on how they can be reduced. The following section makes some suggestions to this effect.

Box 1.4. **Use of Information technology to reduce PRTCs**

Although initial costs can be significant, computer-based technologies help reduce implementation and control costs in the long-term. They facilitate the processing of information and application forms, and the carrying out of automatic controls. Information on programmes, practices or risk management techniques can be accessed on-line. Administrative forms can be made available and processed electronically. Planting and conservation practices can be monitored using GIS (Geographical Information Systems). For example, ASERCA in Mexico uses "spot" GIS system to monitor PROCAMPO. In France, the CNASEA uses a map and information system software (ADAGEO, *Logiciel cartographique et système d'information*) to monitor the implementation of RDR measures, including French land management contracts (CTE). The EU Integrated Administration and Control System uses GIS information to help manage and control land-based and tree-based (olive and nut tree) payments to EU farmers. With 6 million farmers declaring 60 million parcels, 300 000 farms or 3 million parcels are controlled annually.

The use of GIS to monitor and control payments can reduce PRTCs through three channels:

● Digital GIS for field identification reduces error rates and reduces the number of staff required for administrative control. It also reduces the costs of field controls because controls can be targeted. Controllers, who could access all the required information from the same source, are better prepared; and controls can therefore be quicker. The cost of administration is also reduced as filing, archiving and exchange of files are less costly than with printed support and losses are less likely.

● Controls with Remote Sensing have a much lower cost than visit controls and the significant reduction in the number of required visits is less disruptive to farmers.

● Internet access for farmers reduces the distribution costs of maps and forms and the time of staff for such administrative tasks.

Linear programming, simulation models can be used to assist decision-making related to programme options.* In general, administrative and other databases can be used, alone or with modelling tools, to evaluate policies *ex ante* and *ex post*.

* For example, ANAIS (*Logiciel d'analyse micro et d'aide à la décision*) in CNASEA.

Source: Chapter 4; Presentation made at Session 5 of the Workshop on Policy-related Transaction Costs (OECD, 2005a) by Jacques Delincé, Joint Research Centre of the European Commission (*www.oecd.org/agr/meet/prtc*).

Ways to reduce PRTCs

Considering how programme design, institutional systems, farm structures and information systems affect PRTCs (see previous section), the following specific suggestions for reducing the PRTCs of several policy options with similar desired outcomes have been identified:

● Take PRTCs into account when **designing programmes** to optimise value-for-money and the cost-benefit ratio. Trade-offs have to be considered, between precision and implementation costs, or between the level of compliance and the degree of monitoring.

● Programmes can be designed to reduce **information asymmetry** and therefore control costs (when compliance is required). Fraser (2004) suggests using risk aversion to reduce moral hazard and therefore control costs. Increasing monitoring (*i.e.* the probability of being detected) in a target group and increasing monitoring/penalty parameters in the risk-adverse non-target group, would minimize moral hazard. Millock *et al.* (2001)

propose a scheme to regulate pollution that differentiates taxation according to voluntary adoption of monitoring equipment in order to reduce control costs and optimize outcome. Millock *et al.* apply the same model to payments for environmental services and shows that, when there is asymmetric information and verification of actual production practices is costly, monitoring a subset of agents and providing them with higher incentives (or less strict conditions) than non-monitored agents can help reduce control costs and obtain a high level of achievement. Control costs can be reduced without changing compliance with a lower frequency of monitoring but with significant penalties for non-compliance. Some authors suggest selecting programme participants (*i.e.* project proposals) through auctions helps reducing PRTCs while optimizing outcomes (Latacz-Lohmann and Van der Hamsvoort, 1998; Stavins, 1995). Stavins (1995) also suggests that programmes be designed to provide the information needed for monitoring, thus reducing future information costs. Intermediaries possessing information (like brokers in the case of tradeable permits, insurance companies) could help reduce PRTCs if the fees they receive, which are PRTCs, are lower than the gains from the reduction in information asymmetry.

- Limit the number of regulations and do not change them too frequently.
- Make PRTCs **transparent** and monitor them.
- Increase the **duration** of programmes (if proved efficient). Duration is important for two reasons: Set-up costs can be spread over a longer period and some continuity in policy is necessary to take advantage of the accumulated experience of farmers and programme managers. This allows for the rationalisation and automation of procedures, using time-saving technologies. Because of built-in experience, programmes become easier to manage and apply with time. Consensus and co-ordination also improve with time, thus reducing coordination and control costs.
- **Sharing of experience** is also important. It has been suggested that programmes should be implemented incrementally to build on successful experience. Another suggestion is the creation of an independent agency to promote successful experiences. Negative experiences can also be instructive.
- Build on **existing institutions** to implement policies. The development of a common, joint administration network for all policies, for example, has been suggested.
- Reduce the **number of agencies** involved in implementation. For example, farmers would face a single agency/desk for all policies.
- **Contract out** some routine administrative activities through competitive tendering to reduce costs.
- **Increase some PRTCs** to reduce others. For example, activities to promote and explain the scheme, to build a consensus, would reduce implementation and control costs because they increase participation (economies of scale) and compliance (lower control cost).
- Use **information technologies** such as computer databases, on-line information and forms, GIS, cartographic (mapping) software, etc.

While monitoring and reducing PRTCs contributes to government efforts towards administrative efficiency, they also need to take account of PRTCs, together with other costs and benefits, when designing the most economically cost-effective and efficient policy. The following section addresses this issue.

Notes

1. For a discussion on the various definitions of transaction costs in the literature, see Allen (1991).

2. The costs of implementing market price support measures, in particular the domestic measures that often accompany border restrictions, should not be underestimated. The cost of complying with regulations for farmers and the agri-food industry, and the cost of enforcing compliance for the government and inspection agencies are also a concern for the achievement of objectives.

3. The sub-categories of PRTCs incurred in the implementation of voluntary schemes based on compensated management agreements for example are detailed in Table 4.1 of Falconer and Whitby (1999a). It identifies and distinguishes fixed from variable costs borne by the State agency and by participants.

4. Investment costs (buildings and equipment) are usually not considered.

5. Average farm income is used as a proxy for the marginal labour income of farmers. Farmers are assumed to be fully employed in farming.

6. Precision in the definition of policy objectives, using measurable criteria to define the population, area or outcome targeted.

7. Main results are reported in Falconer and Whitby (1999a). Some results are also presented in Falconer (1998). For more information on estimations, see research report FAIR1/CT95-0709/C3/TR8 to the EU.

8. As part of this evaluation, the French authorities included an estimation of PRTCs.

9. Moral hazard refers to the fact that, because of information asymmetry, farmers can receive payments without complying fully to requirements. In the case of insurance programmes, moral hazard occurs when farmers alter their behaviour (i.e. adopt riskier a behaviour) because they are insured.

10. In the case of agri-environmental measures, in particular pollution reducing measures, outcomes are often valued as the cost of compliance (see for example McCann and Easter, 1999, 2000). The most cost-efficient policy is therefore that that achieves the set objectives at the lowest total cost (sum of PRTCs, compliance costs and compensation payments).

11. See OECD (2000a and 2006a) for a discussion on valuation methods.

12. Unit PRTCs depend on the complexity of the task required. The term refers to the cost per transaction and not to the total PRTC as a percentage of transfers.

13. However, it is not always feasible, for example in the case of new EU member countries, which have to build a new implementation framework to adopt EU policies.

14. The issue of property rights in the context of agri-environmental policies is discussed in OECD (2001d).

ISBN 978-92-64-03091-6
The Implementation Costs of Agricultural Policies
© OECD 2007

PART I

Chapter 2

Policy-related Transaction Costs and Policy Choice

2.1. Background

As explained in the introduction, the question of PRTCs in policy choice arises from concerns that moving from direct market interventions to new forms of targeted and decoupled agricultural policies leads to implementation costs that might outweigh the benefits.

This chapter examines the role of PRTCs in defining the most cost-effective and efficient policy option for achieving given policy objectives. Section 2.2 provides a framework for comparing policy options in the context of policy reform that includes a move away from production-linked support towards measures that are more decoupled from commodity production and better targeted to specific policy objectives and beneficiaries. The main economic issues raised when trying to compare policies are highlighted. Scenarios illustrating them are presented. Section 2.3 considers policies that pursue correction of market failures relating to nature, environment and rural viability, while Section 2.4 considers policies with multiple objectives. Finally, agricultural policies with income objectives are considered in Section 2.5.

Because the comparison is being made in a context of policy reform, the costs of different policy alternatives are compared to the costs of pre-reform policies (e.g. market price support). The analysis takes as a starting point that the need for policy intervention has been established and that the objectives of the intervention is well-defined. Once specific, alternative policy measures, which are considered ex ante capable of achieving the objectives, have been identified, which criteria to be taken into account need to be decided in the choice of policies. These include welfare impacts, which can be affected by PRTCs and/or distributional issues. The criteria and the weight given to them can vary depending on the context, social preferences and feasibility.

Annex I.2 contains a brief discussion of marginal cost of taxation issues. Annex I.3 presents a graphical method to illustrate policy comparisons, while Annex I.4 presents comparisons carried out with alternative parameter values.

2.2. Method of comparison

This analysis extends the traditional welfare analytical framework by including PRTCs into cost-benefit analysis. It also includes monetary transfers, which are typically kept outside the analysis in traditional welfare analysis because transfers do not affect overall welfare, only its distribution. Two elements are therefore considered to guide policy choice:

1. welfare changes, including PRTCs; and

2. changes in policy transfers.

To the economist the natural way to evaluate a policy change is in terms of its effects on welfare. This concept attempts to measure in monetary terms whether the society as whole would be better off with the policy change. In general terms, the contemplated policy change would be recommended if it brings a positive net contribution to society's welfare, regardless of the distribution of costs and benefits.[1] Box 2.1 further explains the terminology used. For a

Box 2.1. **The components of welfare changes**

In a well functioning market economy, the current market price reflects the consumers' marginal willingness to pay for a commodity or service. The production of the good requires inputs (labour, intermediate inputs, energy, etc.) that must be withdrawn from the production of other goods and services. The resulting forgone production can be evaluated using the appropriate prices. If input markets are working well, the firm's variable production cost plus any fixed costs, is the appropriate measure of society's valuation of the production forgone in other sectors of the economy (Johansson, 1991).

The welfare effects of market price support measures that raise the prevailing price above its original level can be illustrated in welfare economic terms using Figure 2.1. The area under the demand schedule represents aggregate willingness to pay for the good, while the supply schedule reflects marginal cost of production. Suppose the initial price equals Po, which is the opportunity cost at which this good can be obtained by importing it. Indeed at the price Po demand exceeds supply and the country is a net importer of the good, with imports equal to QDo – QSo.

Figure 2.1. **Graphical illustration of welfare analysis**

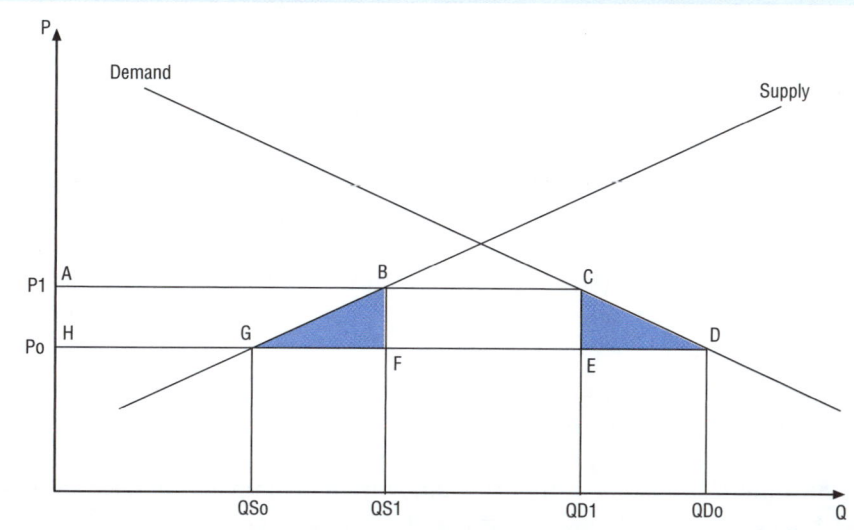

Now consider the effect of a border tariff that raises the domestic price level to P1. Demand for the good drops while producers have an incentive to expand production following the higher price. Producer revenue increases, but variable cost also increases. Netting out both effects, **producer surplus** (the excess of revenue over variable cost) increases by the area ABGH. The new higher price causes **consumer surplus** (the area under the demand curve) to shrink by ACDH. Since the government collects tariffs its revenues increase by BCEF. The net change in welfare equals the sum of these three separate effects. In our case, the decrease in consumer surplus is partly offset by a transfer to producers in terms of an increase in producer surplus, and is partly offset by increased tariff revenues. The additional government revenue may or may not be redistributed in some way. On balance a net welfare loss occurs which in money terms equals the two shaded triangles. This so-called **deadweight loss** from taxation is an uncompensated loss in welfare and represents part of the **resource cost** to society. The triangle BFG informs us about the change in the value of production forgone in other sectors of the economy, when

Box 2.1. **The components of welfare changes** *(cont.)*

moving from the initial situation to the new price-output configuration (P1, QS1). The triangle CDE informs us about the change of the value of foregone consumption on other goods or services as a larger part of the consumer's budget has to be spent on the good under consideration.

- **Outcomes:** They are defined in this study as the results of the policy. They include intended effects with respect to policy objectives (desired outcomes), as well as associated side-effects such as non-commodity outputs and negative externalities. In this study, it is assumed that all the policy options provide the same policy outcomes.

- **Externalities:** They are costs or benefits arising from an economic activity that affect economic agents other than those engaged in the economic activity and are not reflected fully in market prices. Negative externalities are economic costs, while positive externalities are economic benefits. If the production of a good brings with it side effects that affect other parties, and these other parties care about being affected, the social value of the activity diverges from the private value. In the case of a positive externality, when other parties' utility is favourably affected, the social value of production is higher than the private cost. Conversely, in the case of a negative externality, the social value is lower than the private cost of production and it becomes desirable to restrict production to a level where the social cost equals social value at the margin. In terms of Figure 2.1 this would require the inclusion of **social marginal cost** schedules instead of the private schedules depicted (see Figure 2.3 of Box 2.4). Negative externalities of agricultural production include nitrogen and pesticide run-offs, green house gas emissions or bad smells.

- **Non-commodity outputs:** The term as used here was elaborated in the context of OECD's work on multifunctionality (OECD, 2003a, Box 1). It relates to a wide range of positive effects of agriculture whose relationship to agricultural production is described as the degree of jointness (Box 2.3). Rural viability, biodiversity and landscape have, *inter alia*, been claimed as positive effects or non-commodity outputs of agriculture. In practice, jointness may be weak or strong.

- **Additional cost of de-linkage (K):** In the context of policies that aim to correct market failures, this analysis takes account of the total cost of producing a non-commodity output separately from commodity production by adding the additional cost of producing a non-commodity output separately from commodity production (K) to the intended transfer needed to produce it jointly (Y). K is in fact equal to the total cost of producing the public good separately, minus the intended transfer (Y) needed to produce it jointly with the commodity output, using a targeted coupled policy. This difference in economic cost is assumed to be reflected in the level of transfers needed to obtain the output. K is envisaged to be measured as the difference in intended transfers between a coupled policy (joint production) and a decoupled policy (separate production). An underlying assumption for this is that with separate production it is possible to target the transfers perfectly to compensate for the costs of producing the public good. If a coupled policy transfers X (or Y), the equivalent decoupled policy will transfer X + K (or Y + K). K is positive if joint production is cheaper (in terms of intended transfers), and negative otherwise. When K is positive, it can vary from zero to infinity. Since all policy options are assumed to provide the same policy outcome, K represents the additional resources consumed while not producing anything additional to the non-commodity output. Therefore, K is regarded as a part of the resource cost.

- **PRTCs:** As the additional cost of de-linkage, PRTCs represent resources consumed and are therefore considered as a part of the resource cost.

complete evaluation of the effects of a policy change it is important to include all the relevant effects associated with the change, and to value the effects using the appropriate prices. In all cases, the welfare analysis attempts to measure in monetary terms the opportunity cost of carrying out the particular policy change (or commencing a particular project). In other words, it poses the question "what is the value of production and consumption foregone in other parts of society if we carry out the particular policy change?" In the applied version of welfare economics – traditional economic cost-benefit analysis – this translates to using border prices to value importable commodities in order to approximate society's opportunity cost of consuming that particular good. This is also the method adopted by the OECD to evaluate the market price support component of the Producer Support Estimate (PSE).

First, a series of costs and benefits that affect overall welfare are considered. They are deadweight losses on the production and consumption side (DWL), possible additional costs of de-linkage (K) due to decoupling in the context of market failures, when jointness exists (i.e. the extra cost of producing a non-commodity output separately from commodity production – to be added to the transfer to producers needed to produce it jointly – see OECD, 2001a and 2003a, and Box 2.1), marginal cost of taxation (MCT), the outcome of the policy (OUT) and associated changes in the value of positive and negative externalities that are not intended by the policy (ΔEXT). These elements are further defined in Box 2.1.

The move from policy i to policy j will be recommended if:

$$(DWLj + Kj + MTCj + OUTj + \Delta EXTj) - (DWLi + Ki + MCTi + OUTi + \Delta EXTi) < 0 \qquad [1]$$

The marginal cost of taxation is not considered numerically in the policy comparison (i.e. MTCi is not considered further). Annex I.2 contains a brief discussion on MCT issues and reports some estimates of the MCT of public funds. They vary widely by country as they depend on the composition and level of taxation.

Moreover, because the value of all costs and benefits cannot be estimated and for the sake of simplification, this analysis compares policies that are assumed to achieve the same desired outcome (i.e. outcome with respect to the objective set). In other words, OUTi = OUTj. The existence of positive and negative externalities (other than the one targeted by the policy) is acknowledged but not quantified given the theoretical and empirical difficulties in evaluating those effects. Thus ΔEXT is not considered further.

Second, changes in **PRTCs** associated with the policy change are taken into account. PRTCs (PRTCi) can also be measured in monetary terms, using methods described in Section 1.4. Conceptually, a complete cost-benefit evaluation of a policy change can be written as the difference of the conventional change in welfare associated with the policy change minus the change in PRTCs needed to implement the policy change. Given the assumptions retained in this analysis, policy j is now to be preferred to policy i if:

$$(DWLj + Kj + PRTCj) - (DWLi + Ki + PRTCi) < 0 \qquad [2]$$

Third, transfers within the economy are considered.[2] These transfers are not considered explicitly in traditional cost-benefit analysis (other than through their impact on the size of costs and benefits) because they do not affect overall welfare. However, they are important for society because they affect the distribution of welfare among households or sectors. They may result in inequities between households and sectors that are a concern for policy-makers and society overall. They are also a concern for policy makers, in particular in the context of sectoral policy reform, because they may cause financial waste if transfers are greater than needed to meet the desired objective or if they leak to unintended beneficiaries (Box 2.2). It is all the more important to include unintended transfers in this context as one objective of policy reform is to obtain desired outcomes while minimising government expenditures. How much society is prepared to pay is therefore an important component of policy choice. This approach is consistent with the long established measurement of transfers to agriculture that is undertaken in the form of the OECD's Producer Support Estimate (PSE) and Consumer Support Estimate (CSE). These estimates track transfers from consumers in the form of market price support calculations and from taxpayers in the form of direct budgetary payments to farmers. Both are considered as legitimate elements to be accounted for in the costs of policies. This analysis therefore also includes transfers in the comparative equations that are used to evaluate policy alternatives.

Box 2.2. **Targeting concept**

- **Intended transfers:** They are the transfers to agricultural producers that are needed to produce the desired outcome, and only those transfers.

- **Unintended transfers:** They are transfers to agricultural producers that do not produce the desired outcome, either because they go to unintended recipients, or because they are higher than needed to produce the desired outcome. In other words, they are not necessary to achieve policy objectives, but exist because the policy is not well targeted or tailored.[1]

- **Targeting**[2] requires well defined objectives that allow the elements to receive support (population, area or outcomes) to be clearly identified and the level of support required to be specific to the objectives. This also means the policy is tailored, i.e. provides only the amount of transfers needed to obtain the outcome, a notion often associated with good targeting. As a result of good targeting and tailoring, transfers needed to achieve objectives are expected to be lower than transfers from broad-based policies. This assumes, of course, that targeting is technically possible.

- **Targeting ratio:** This is the ratio between the transfers needed to meet the objective (using a targeted policy: Y) and the transfers from a broad-based policy (X) that achieves the same objective.

1. It should be noted that targeting may not always be technically feasible and/or may raise equity concerns. One example of the latter is when payments are made to farmers to change their production practices and thus produce additional non-commodity outputs, but farmers who already produce the non-commodity output do not receive any payment for doing so.
2. The notion of targeting is being further elaborated as part of the OECD programme of work and budget for 2005-06. The project seeks to draw lessons on best practices for effective targeting.

Agricultural policy traditionally delivered transfers to farmers through broad-based measures such as market price support or output payments and a significant share of transfers to the sector are still delivered in this way (OECD, 2005c). In recent years there has been a move towards area or animal payments. This type of measure is often also broad-based in the sense that all cropland, all land or all animals receive the payment, although increasingly, there are restrictions on the number of hectares or animals to receive support. If the objective of the policy is to raise the income of farmers to some minimum level, a broad policy such as price support or a commodity-specific payment will generate transfers to producers who already have the minimum income. To the extent that this happens the transfers are unintended. Other examples of unintended transfers would be area payments for the maintenance of farming in places where farming would be viable without it; or payments for the adoption of production practices favourable to biodiversity in areas where the desired level of biodiversity is already achieved. On the other hand, a targeted payment, as opposed to a broad based measure, will reach only those farmers who need it – in the case of a minimum income goal – or only those farmers who increase supply – in the case of a positive externality. These concepts of unintended transfers and of targeting are explained more formally in Box 2.2.

It follows therefore that transfers aimed at producers (T) can be broken down as follows (Figure 2.2):

1. intended transfers (Y) that reach intended beneficiaries and only provide the amount necessary to obtain the desired outcome (Box 2.2);

2. unintended transfers (Z) that are not needed to produce the desired outcome (Box 2.2);

3. deadweight losses (DWL) (Box 2.1); and

4. additional costs of de-linkage (K) (Box 2.1).

Figure 2.2. **Relationships between economic resources and transfers**

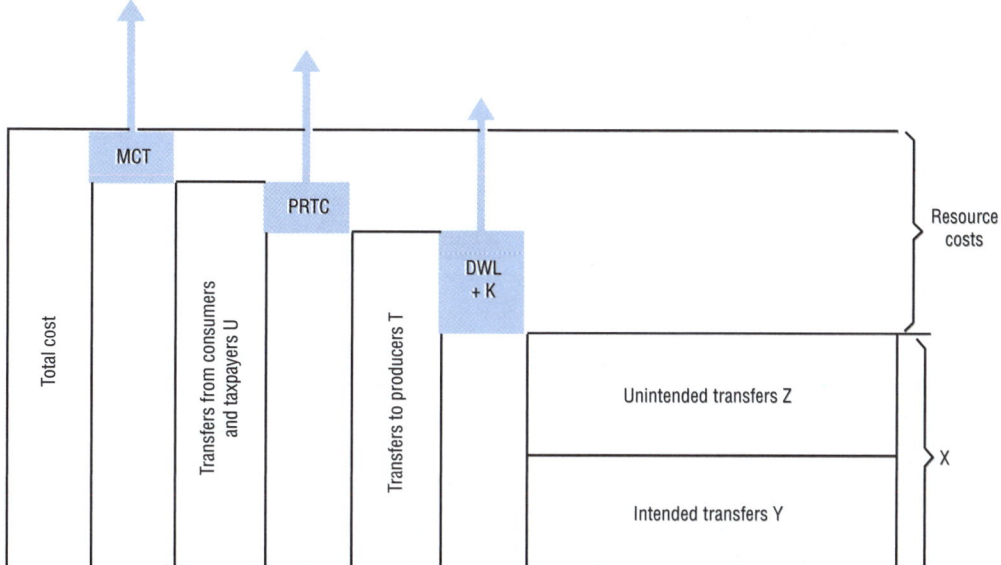

The size of the different blocks does not reflect the size of the different elements.
MCT: Marginal cost of taxation; DWL: Deadweight losses; K: Additional cost of de-linkage.
Transfers from a targeted policy are Y, while transfers from a broad based policy are X = Y + Z.
Source: OECD Secretariat.

A broad-based, coupled policy will transfer T = Y + Z + DWL, with producers receiving X = Y + Z (Figure 2.2). A targeted, decoupled policy will only transfer the amount necessary to obtain the desired outcome and there will be no unintended transfer and no deadweight losses (T = Y or T = Y + K).

The relationship between the economic resources needed to pursue a given objective and the transfers received by farmers, given the various costs incurred, is illustrated in Figure 2.2. If the policy involves budgetary expenditures, there is a cost to the economy in raising taxes (the marginal cost of taxation, MTC) so the economic cost is higher than transfers from consumers and taxpayers. Those are then partly used to fund administrative costs (PRTCs) and the remaining transfers from consumers and taxpayers are aimed at producers (T). Some transfers then do not reach farmers because of deadweight losses (DWL).[3] Transfers received by farmers are finally divided into intended transfers (Y) that generate the desired outcome, and unintended transfers (Z) that occur either because recipients do not generate the desired output or do so above the level needed (*e.g.* in an area where it is not needed).

So far, it has been established that two elements should be retained in order to compare the costs of different policies that are assumed to bring about the same outcome in terms of the objective sought. These are resource (or welfare) costs, composed of deadweight losses, the additional costs of de-linkage and PRTCs, and (intended and unintended) transfers. Welfare elements relate to economic efficiency, while transfers relate to distribution. Policy-makers may wish to consider these aspects separately, as illustrated in Figure I.3.1.

In Figure I.3.1, resource costs are represented on the X-Axis and unintended transfers on the Y-Axis. Policy O is inferior to Policy A while Policy B is inferior to Policy O. However, we cannot say whether any policy falling in the grey areas is inferior or superior to Policy O. When the choice is indeterminate, policy-makers might want to weigh the two types of costs.[4] While they might wish to give different weights to the two types of costs or to any of the individual costs, benefits and transfers,[5] this analysis assumes a dollar of welfare gain is equivalent to a dollar of welfare loss or to a dollar of transfer, whoever is affected. This assumption is made in the absence of any plausible alternative. The move from policy i to policy j then depends on whether:

$$(DWL_j + K_j + PRTC_j + Y_j + Z_j) - (DWL_i + K_i + PRTC_i + Y_i + Z_i) < 0 \qquad [3]$$

Where DWL_i are the deadweight losses, K_i the additional cost of de-linkage, $PRTC_i$ the policy-related transaction costs, Y_i intended transfers and Z_i unintended transfers of policy i.

This simplified comparative approach is more practical than a standard welfare maximisation approach, where all non-commodity outputs and negative externalities generated by the different options would have to be valued simultaneously. Essentially the approach is a cost-effectiveness analysis, which takes the outcome as given and ranks policies according to the total cost needed to achieve the outcome. This method is preferred, both because the nature of the changes in many of the variables (positive and negative externalities) and their value are not known, but also because the change in the level of transfers is a legitimate and potentially important element in policy choice, reflecting decisions about who should pay and how much. The alternative method chosen here allows policy options to be compared, taking account both of transfers and those economic costs that are measurable.[6]

2.3. Application to policies aiming to correct market failures

This section looks at how the different cost elements vary with different policy options, in the case of a policy intervention that is designed to deal with a market failure. Possible examples of market failure could relate to the provision of landscape or of biodiversity. The different policies considered include market price support, and payments with different combinations of full or no targeting and decoupling. Payments are either fully coupled or fully decoupled, and/or perfectly targeted or untargeted. In reality, there are few examples of fully decoupled policies as any policy based on farm resources is likely to have some impact on farm production (OECD, 2001f). Similarly, examples of perfectly targeted policies are rare.

Jointness

The necessity to compare policy options that range from market price support – a broad-based policy – to a targeted, decoupled instrument, derives from the work on multifunctionality. It relates to the debate concerning the degree of jointness – the nature of the link between commodity production and the non-commodity outputs of agriculture (Some technical definitions of jointness are given in Box 2.3). If agricultural production is strongly related to the desired non-commodity outputs – if production of the non-commodity output is strongly correlated with increases in production of the commodity – a case can be made for a broad-based policy that encourages commodity production in order to produce the desired non-commodity output. If, on the other hand, jointness is weak, the case for decoupled and targeted policy instruments is more likely to prevail. This is the underlying policy choice that the following paragraphs illustrate.

Box 2.3. Jointness and related concepts

- **Jointness:** Joint production refers to situations where a firm produces two or more outputs that are interlinked so that an increase or decrease of the supply of one output affects the levels of the others. Three reasons for jointness are frequently distinguishes: 1) technical interdependencies in the production process (*e.g.* crop production and nutrient leaching); 2) non-allocable inputs (*e.g.* sheep raising producing mutton and wool; production systems and associated landscape); and/or 3) allocable inputs that are fixed at the firm level such as farmland and labour (OECD, 2001a, Box I.5). Joint outputs of agriculture vary from private to public goods, and include goods with various degrees of public good characteristics. Similarly, they display various degrees of jointness with commodity production, depending on the extent to which the share of the various joint outputs can be modified.

- **Economies of scope:** They are possible cost savings due to joint production. Economies of scope arise if something inherent in the production process makes it cheaper to provide two or more outputs jointly rather than separately.

- **Jointness and deadweight losses:** When transfers are used to produce a non-commodity output jointly with the commodity output, the net change in producer and consumer surplus is partially or totally offset in the welfare equation by the value of the non-commodity output generated. However, deadweight losses used for the additional commodity production are retained in equation [3] because the output side (*i.e.* the value of the desired non-commodity output) cancels out on both sides of the equation as all policies are assumed to have the same outcome. Moreover, when the non-commodity output can be produced either jointly or separately, the resource used in the case of joint production leads to additional commodity production that is not needed, while this could be avoided if a decoupled policy was adopted. The deadweight losses are therefore attributed to the commodity output, and therefore to the coupled policy.

Not all comparisons are possible or meaningful. Coupled and decoupled options can be compared only when both joint and separate production are technically feasible. If no separate provision is possible, the decoupled option will not meet the criteria of achieving the same outcome. Moreover, there are cases where the choice is obvious and does not require a detailed analysis. For example, if the total transfers needed to produce the desired outcomes with a targeted policy are higher than transfers from a broad-based policy that generates the same outcomes, the latter is to be preferred, as PRTCs are also likely to be lower.

In other cases, the choice is not immediately obvious, as the desired outcome can be obtained using either a policy coupled to commodity production (market price support or coupled payment) or a policy that is not linked to commodity production (decoupled payment) if jointness is not perfect and de-linkage feasible.[7] For example, flood control can be pursued by raising the rice price, paying for hectares of terraced paddy field or building a dam. Both support to grazing livestock, pastures, or hedges can contribute to biodiversity improvements if appropriate conditions are attached. Obtaining the desired outcome separately from commodity production may be more expensive. This is the case if there are economies of scope, *i.e.* if the production of the non-commodity outputs jointly with the commodity output is cheaper than separate production (Box 2.3). Conversely, there will be a benefit if separate production is cheaper.

Cost comparison

Price support results in deadweight losses (welfare triangles) and world price distortion. The same deadweight losses (excluding the consumer welfare triangle and with somewhat smaller world price distortions) are found with coupled payments. With perfectly decoupled payments, deadweight losses are zero. Total deadweight losses from policy i (DWLi) are equal to per unit deadweight losses (dwli) multiplied by the total transfer.[8] PRTCs per unit of transfers[9] are labelled tci and total PRTCs PRTCi for policy i. As in equation [3] and Figure 2.2, the intended transfer, achievable through a well targeted policy, is Y, while the actual transfer, resulting from an untargeted policy, is X, with X > Y (Z = X − Y are unintended transfers). For decoupled policies, the additional cost of de-linkage (K) is added. The additional costs of de-linkage per unit of transfer (X or Y) are labelled ki and the total additional costs of de-linkage Ki for policy i.

Table 2.1 allows for a formal comparison of cost elements. Policies can be compared two by two for all costs retained in equation [3], *i.e.* resource costs (deadweight losses, additional cost of de-linkage and PRTCs) and transfers.

Table 2.1. **Market failure: Comparison of costs by policy type**

For a given outcome	Price support (tariff only)	Price support (complex trade and domestic measures)	Untargeted coupled payment (*e.g.* output payment)	Untargeted decoupled payment (*e.g.* pay. based on historical entitlement)	Targeted coupled payment (*e.g.* limited output payment)	Targeted decoupled payment (*e.g.* targeted payment per meter of hedge)	Policy with no transfer (*e.g.* regulation)
Policy *i*	1	2	3	4	5	6	7
Intended and unintended transfers Xi	X	X	X	X	Y	Y	0
of which: unintended transfers Zi	Z	Z	Z	Z	0	0	0
Total additional cost of de-linkage Ki	0	0	0	k4*X	0	k6 * Y	0
Total deadweight losses DWLi[1]	dwl1 * X	dwl2 * X	dwl3 * X	0	dwl5 * Y	0	DWL7
Total PRTCs PRTCi	tc1 * X	tc2 * X	tc3 * X	tc4 * (X + k4 * X)	tc5 * Y	tc6 * (Y + k6 * Y)	PRTC7
Transfers to producers Ti	X + dwl1 * X	X + dwl2 * X	X + dwl3 * X	X + k4 * X	Y + dwl5 * Y	Y + k6 * Y	DWL7
Total cost for consumers and taxpayers Ui	X + dwl1 * X + tc1 * X	X + dwl2 * X + tc2 * X	X + dwl3 * X + tc3 * X	X + k4 * X + tc4 * (X + k4 * X)	Y + dwl5 * Y + tc5 * Y	Y + k6 * Y + tc6 * (Y + k6 * Y)	DWL7 + PRTC7

X = transfers from an untargeted policy (X = Y + Z). Y = transfers from a targeted policy; Z = transfers to unintended beneficiaries because of lack of targeting.
1. Ki are the additional cost of de-linkage per unit of transfers; k4 = k6.
2. DWLi are deadweight losses per unit of transfers; dwl1 = dwl2 > dwl3 = dwl5.
Source: OECD Secretariat.

Box 2.4. **Main assumptions on parameters retained to illustrate the comparison**

- **PRTCs:** Plausible assumptions on PRTCs as a percentage of transfers are made, based on estimates found in the literature and the case studies. The unit PRTCs used are those presented in the middle column of Table 2.2.

Table 2.2. **Plausible range of PRTCs as a percentage of transfers by policy type**

Policy	Minimum	Base value	Maximum
MPS tariff only	0.25	0.5	1
MPS other measures	0.44	10	12
Untargeted, coupled payment	1	3	7
Untargeted, decoupled payment	1	3	7
Targeted, coupled payment	2.5	25	50
Targeted, decoupled payment	5	50	110

Source: Secretariat's assumptions based on Tables I.1.2 to I.1.19 and Table 1.3.

- **Deadweight losses:** Table 2.3 contains examples of deadweight losses per unit of transfer for the schematic measures considered. The deadweight losses per unit of transfer associated with MPS and output payments were estimated using the PEM crop model (OECD, 2001c) for different types of crop policies. This estimation refers to resources used to produce commodity outputs. It does not take account of positive and negative externalities other than the ones targeted by the policy.

Table 2.3. **Plausible base values of impacts by support measure Welfare gain or loss per unit of transfer**

For a given outcome	Price support[1]	Targeted or untargeted coupled payment (*e.g.* output payment)[2]	Targeted or untargeted decoupled payment (*e.g.* income payment)[3]
Tax payers	−0.56	−1.31	−1
Consumers	−0.44	0.31	0
Farm households			1
– Land	0.26	0.27	
– Other farm owned	0.14	0.15	
Input suppliers	0.26	0.27	0
Resource costs	−0.34	−0.31	0
Income transfer efficiency ratio[4]	0.27	0.285	1

These numbers are illustrative.
1. Derived from simulating a 5% increase in market price support for wheat in the European Union.
2. Derived from simulating a 5% increase in output payments for wheat in the European Union.
3. Assuming total decoupling.
4. Assuming farmers own 50% of the land they farm.
Source: OECD (2001c), adapted from Table A1.12.

When a coupled, targeted policy is used to correct a market failure in the provision of a positive externality (case of perfect jointness and of separate production not possible), the producer price is raised from Po to P1 in order to reach the social optimum (QS1 in Figure 2.3) and there is no deadweight loss on the production side. If an output payment is used, there is no deadweight loss either on the demand side, but with price support, the deadweight losses remain on the demand side. In addition, if a broad-based policy is implemented, unintended transfers from lack of targeting (i.e. generating a positive externality

> ### Box 2.4. **Main assumptions on parameters retained to illustrate the comparison** (*cont.*)
>
> in places where it is not needed) continue to generate deadweight losses on the production side, even though these are smaller. As explained in Box 2.3, if jointness is imperfect and if separate production is possible, deadweight losses on the production side of targeted and untargeted, coupled policies are attributed to commodity production and continue to be taken into account.
>
> #### Figure 2.3. **Graphical illustration of deadweight losses in the case of joint production**
>
>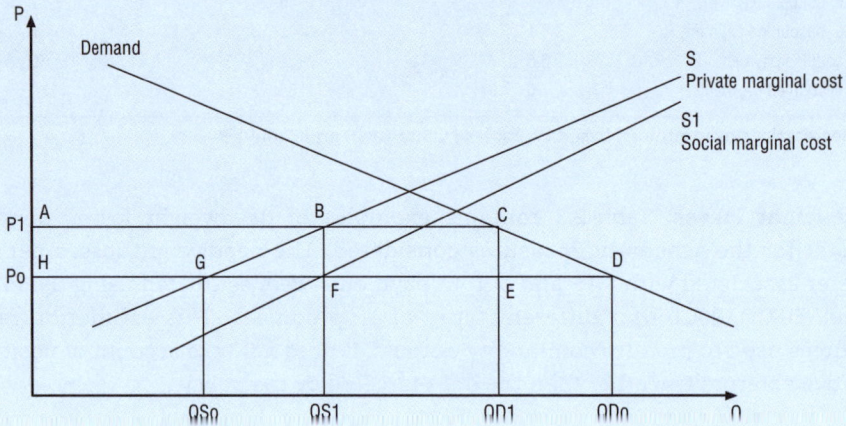
>
> - **Additional cost of delinkage:** We have no information about the value of such costs. For convenience, they are expressed in the illustrative examples as a proportion of transfers (X or Y), with three alternative assumptions. The additional cost of de-linkage is either 0, 20% or 50% of transfers.
>
> - **Targeting ratios:** The estimation of gains from targeting requires structural or regional information. For illustrative, pedagogical purposes, various assumptions are made on the required degree of targeting to achieve the objective.

Numerical examples of the formulas presented in Table 2.1 are presented in Figures I.4.1 and I.4.2. The three examples illustrated in these figures correspond to different situations from strong to weak jointness and from widespread to limited incidence of market failure. These examples are purely illustrative and do not reflect real-life situations. In fact, OECD work on multifunctionality has shown that many non-commodity outputs of agriculture are "specific to a particular site, locality or region" and that "it is not very common for them to be associated with all agricultural production in a country or all land in agricultural production" (OECD, 2003a, p. 70). In these cases, targeting ratios would be low. Non-commodity outputs display various degrees of jointness, which can often be modified by changing production practices. More efforts should be pursued to evaluate potential economies of scope and costs of de-linkage. In particular, the illustrative comparison in this analysis would benefit from improved assumptions on the additional cost of de-linkage. The OECD is contributing to better understanding and measurement of jointness with the organisation of a Workshop on 30 November-1 December 2006.

There are various ways in which the parameters of the policy choices being faced can be presented. For example, for a range of targeting ratios, and a range of PRTCs as a percentage of transfers (%PRTCs) for a "reference" policy, it is possible to estimate the level of the %PRTC of an alternative policy, above which the alternative will prove more costly. This is done in Table 2.4 where a targeted decoupled policy (for example a payment for the maintenance of the agricultural landscape) is compared to the "reference" untargeted, coupled policy (for example, a price support granted with the same objective in mind). In this illustration the same base level of deadweight losses due to the coupled nature of the policy is used as in previous examples. Thus, if the %PRTC of price support is 1, the %PRTC of the targeted, decoupled option could be as high as 1 250 (at a targeting ratio of 0.1) and 50 (at a targeting ratio of 0.9), for the alternative measure still to be preferred. Similarly, if the %PRTC of price support is 50, the targeted, decoupled option is still preferable at a %PRTC of 1 740 (if the targeting ratio is 0.1) and 104, (if it is 0.9). This exercise is repeated for alternative pairings of policy options and presented in detail in Tables I.4.1 and I.4.2, where the impact of variations in the level of deadweight losses associated with the different options is also shown.

Table 2.4. **Market failure: The choice between a targeted, decoupled policy (6) and an untargeted, coupled policy (1) given assumptions on %PRTCs and targeting ratios**

Maximum value of %PRTC (tc6) for the targeted option to be lower cost (%) (illustrative purpose only)

PRTC of the untargeted, coupled policy as a % of transfers (tc1)	Targeting ratio (Y + K)/X										
	Base assumption on the deadweight losses of the untargeted, coupled policy (dwl1 = 0.34)										
	0.1	0.2	0.25	0.3	0.4	0.5	0.6	0.7	0.8	0.9	1
1	1 250	575	440	350	238	170	125	93	69	50	35
5	1 290	595	456	363	248	178	132	99	74	54	39
10	1 340	620	476	380	260	188	140	106	80	60	44
20	1 440	670	516	413	285	208	157	120	93	71	54
40	1 640	770	596	480	335	248	190	149	118	93	74
50	1 740	820	636	513	360	268	207	163	130	104	84

[tc6 * (Y + K) = X − Y − K + tc1 * X + dwl1 * x], thus tc6 in % = 100 * {[1 + tc1/100 + dwl1]/[(Y + K)/X] − 1}.
X = transfers from an untargeted policy; Y = transfers from a targeted policy; K = additional cost of de linkage.
Source: Based on formulas in Table 2.1.

Calculations such as those could be undertaken systematically for real case policies and would permit policy makers to identify where key trade-offs occur. For example, a government faced with a choice between an output payment and a regional area payment, having at its disposal enough information to estimate the targeting ratio and the relative levels of PRTCs, and assuming that the deadweight losses associated with the regional payment would be lower, would be able to make an informed choice in favour of the lower cost policy.

2.4. Application to policies with multiple objectives

In the context of multifunctionality, the issue of PRTCs was raised in terms of the choice between a multiple objective policy that supports commodity production in order to obtain non-commodity outputs, and several policies that would address individually each of the non-commodity outputs. In the comparative analysis presented above, policies are compared two by two. Similar comparative tables could also be developed to compare the combined cost of several individual policies and the cost of one policy with multiple objectives.

A broad-based policy pursuing two objectives can be compared with two targeted policies aiming each at only one of the objectives. For example, support to grazing livestock may generate a landscape amenity and employment on the farm. This type of coupled support can be compared to a regional area payment to keep land open, with production not required, combined with measures to promote employment from rural tourism. Let's assume that the broad-based policy generates deadweight losses of 34% of transfers, and that changes in other positive and negative externalities caused by the policy are offsetting. Transfers are X0. If PRTCs of tc0 are attributed to the first objective, there is no PRTC for the second objective. The first targeted policy transfers X1 (including the additional cost of de-linkage K1) with unit PRTCs of tc1, while the second targeted policy transfers X2 (including the additional cost of de-linkage K2), with unit PRTCs of tc2. Both targeted policies are decoupled from commodity production and have deadweight losses per unit of transfer of zero. In this context the sum of transfers from the two targeted policies (X1 + X2) should be equal or lower than the transfers from the broad-based policies (X0) to justify the choice of the targeted policies. Two targeted policies (1 and 2) are to be preferred to one broad-based policy (o) if:

$$DWL1 + DWL2 + K1 + K2 + PRTC1 + PRTC2 + X1 + X2 < DWLo + Ko + PRTCo + Xo$$

with

$$X1 + X2 < Xo.$$

Many combinations of degree of targeting and unit PRTCs are possible. To illustrate the trade-off between targeting and PRTCs, Table I.4.3 uses various combinations of these parameters and compares total transfers, total deadweight losses and total PRTCs of the two options. It shows that as long as the unit PRTCs of the two targeted measures are within the range of 20-30%, which can be the case according to the literature, they are to be preferred whatever the targeting ratios, as long as total transfers of the two targeted options are lower than these of the broad-based policy. Moreover, when the savings in transfers from targeting are over 25% [(X1 + X2)/X0 < 75%], the combination of targeted policies is most likely to be the lower cost option, even if the PRTCs account for a relatively large share of transfers (up to 50%). However, when the PRTCs of two targeted measures are relatively high (e.g. over 50%), a coupled, broad-based option with low PRTCs might be preferred if the targeting ratio is high. Figure 2.4 shows the minimum degree of targeting necessary for the targeted options to have lower costs relative to the unit PRTCs of the targeted policies (assumed to be equal).

With a multiple objective policy, additional factors should, however, be considered. First, the desired outcome of a broad production linked policy compared to a targeted policy will be the same only if there is perfect jointness between the commodity output and all the non-commodity outputs. A broad-based measure, however, is likely to cause more changes in other positive and negative externalities than a targeted measure. Second, it is likely that the combined PRTCs of policies implemented using the same delivery network and same database are less than if each policy was implemented alone. Third, monitoring and evaluation costs of a multiple objective policy will be higher than

Figure 2.4. **Market failure: Trade-off between targeting ratio and unit PRTCs based on different hypothetical combinations of key parameters**

Minimum ratio (X1 + X2)/X0 for the targeted option to have lower costs (illustrative purpose only)

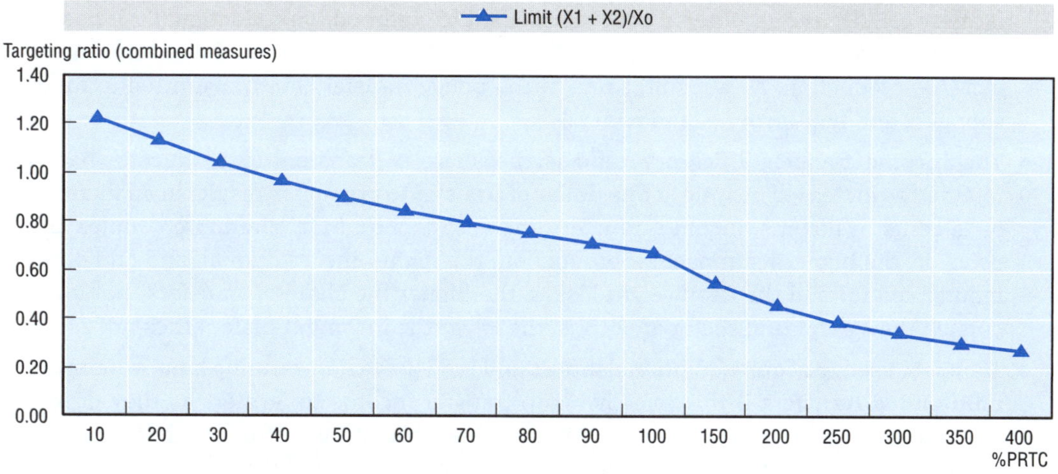

those of a single objective policy but will not necessarily be higher than when several instruments are used in pursuit of several objectives, because several outcomes have to be assessed. Finally, the cost of evaluating the existence and degree of jointness is likely to be high. Another difficulty is that the presence of positive and negative externalities modifies the optimum solution and the amount of deadweight losses per unit of transfers. As a result, the change in deadweight losses when moving from a coupled policy to a decoupled policy is even more difficult to estimate as the shadow prices of all positive and negative externalities need to be evaluated.

2.5. Application to policies with income objectives

In this section, it is assumed that the purpose of the policy intervention is to ensure that all farmers reach a specific minimum income level. The same range of policy interventions is compared using the same methods as in the sections dealing with market failure, but with one major difference. When a coupled policy (*i.e.* an intervention that stimulates production) is used to increase income there are significant leakages to others who may be input suppliers or landowners who do not themselves farm the land. It follows that the transfer that is needed (for example in the form of higher prices paid by consumers) will be greater than the benefit to the farmer in terms of increased net income. A further layer of unintended transfers – those that leak out of the sector – must therefore be taken into account when comparing coupled and decoupled policies whose aim is to increase the income of farmers. For example a general finding is that 4 dollars of market price support will increase net farm income by approximately 1 dollar (OECD, 2001c, 2003b). For the comparisons that follow, the implication is that for farmers to receive an income boost of X through price support, the initial transfer may need to be as much as four times greater. This concept of income transfer efficiency is explained more formally in Box 2.5.

Estimates of the transfer efficiency per unit of transfer of various policies are derived from previous OECD work (OECD, 2001c) and presented in Table 2.3 of Box 2.4. The income transfer efficiency of market price support is estimated to be 27%, *i.e.* that 27% of transfers end up as additional net farm income ($X = 0.27 * T$), remuneration of labour and land) and

---

Box 2.5. **Income transfer efficiency concepts**

- **Income transfer efficiency** measures the efficiency of policies at transferring income to farmers (OECD, 2003b). When a policy is coupled to commodity production (*i.e.* it has an impact on production), producers purchase/rent additional variable inputs and land to increase production. As a result, some of the policy transfers are transmitted to input suppliers and non-farming landowners.

- **The income transfer efficiency ratio** is the share of transfers to producers that is retained as increased income. If one dollar of transfer to producers results in an increase of 25 cents in income, income transfer efficiency is a quarter. The more coupled the policy is, the higher its impact on production, the higher the additional land and other inputs paid for and the deadweight losses, the higher the share of transfers that leaks outside the agricultural sector and, thus, the lower the income transfer efficiency ratio.

- Transfers used to acquire additional inputs and land (from non-farming landowners) are **unintended transfers** if the objective is to transfer income to farmers as they do not contribute to the outcome. They leak outside the agricultural sector and benefit input suppliers and non farming landowners, who are not the target of the policy (unintended beneficiaries). Therefore, when the objective of the policy is to transfer income to farmers, they need to be taken into account in addition to resource costs and unintended transfers within the agricultural sector already mentioned. When the objective of the policy is not to transfer income to farmers but to obtain non-commodity outputs, the fact that part of the transfer leaks from the agricultural sector is not an issue if it is used to obtain the desired outcome. Income transfer efficiency is therefore not considered other than in examples relating to income support.

---

thus almost three quarters goes to unintended beneficiaries outside the sector such as non-farming landowners (nfl = 13%) and input suppliers (is = 26%) or are lost because of distortions in the allocation of resources (dwl = 34%). In equation form:

$T = DWL + NFL + IS + X$ or

$T = dwl * T + nfl * T + is * T + X$

It follows that $X = T * (1 - dwl - nfl - is)$ and $T = X/(1 - dwl - nlf - is)$

if dwl = 0.34, nfl = 0.13 and is = 0.26 as in the case of price support, then

$T = X/(1 - 0.34 - 0.13 - 0.26) = X/0.27$

This means that for transfer X to reach producers, transfers from coupled payments need to be equal to T = X/0.27. In the case of a decoupled payment that is not linked to land, transfer efficiency is 100%.

The derivation of total initial transfers by type of measure is presented in Table 2.5. Table 2.6 then compares unintended transfers, deadweight losses and total PRTCs for the different measures considered.

Figure 2.5 illustrates the difference between total costs of different policy options. All decoupled options prove less costly than coupled instruments, reflecting the significance of both the leakages and the deadweight losses associated with this type of policy. A targeted decoupled policy is the least-cost solution but even an untargeted, decoupled measure is significantly less costly than any coupled option.

Table 2.5. **Derivation of the total transfer necessary to increase income by Y using assumptions on key parameters**

Illustrative purpose only

For a given outcome (*e.g.* minimum income level)	Price support	Untargeted coupled payment (*e.g.* output payment)	Untargeted decoupled payment[2] (*e.g.* pay. based on historical entitlement)	Targeted coupled payment (*e.g.* limited output payment)	Targeted decoupled payment (*e.g.* targeted income payment)
Policy i	1	2	3	4	5
Transferred income to farmers X_i	$X = 0.27 * T1$	$X = 0.285 * T2$	$X = T3$	$Y = 0.285 * T4$	$Y = T5$
Deadweight losses DWL_i	$0.34 * T1$	$0.31 * T2$	0	$0.31 * T4$	0
Unintended transfers from leakages to landowners and input suppliers W_i	$(1 - 0.27 - 0.34) * T1$	$(1 - 0.285 - 0.31) * T2$	0	$(1 - 0.285 - 0.31) * T4$	0
Original transfers T_i[1]	$T1 = X/0.27$	$T2 = X/0.285$	$T3 = X$	$T4 = Y/0.285$	$T5 = Y$

X = transfers from an untargeted policy; Y = transfers from a targeted policy; Z = transfers to unintended beneficiaries because of lack of targeting; W = transfers to unintended beneficiaries because of transfer inefficiency.
1. $X = r * T$ implies $T = X/a$ with "r" being the transfer efficiency ratio. T is also equal to $Y + Z + W + DWL$.
Source: OECD Secretariat.

Table 2.6. **Income support: Comparison of costs by policy type, using assumptions on key parameters**

Illustrative purpose only

For a given outcome (*e.g.* minimum income level)	Price support	Untargeted coupled payment (*e.g.* output payment)	Untargeted decoupled payment[1] (*e.g.* pay. based on historical entitlement)	Targeted coupled payment (*e.g.* limited output payment)	Targeted decoupled payment (*e.g.* targeted income payment)
Policy i	1	2	3	4	5
Transfers T_i	$X/0.27$	$X/0.285$	X	$Y/0.285$	Y
Unintended transfers					
– Within the sector (from lack of targeting) Z_i	Z	Z	Z	0	0
– Leakages to other sectors (transfer efficiency losses) W_i	$0.39 * X/0.27$	$0.405 * X/0.285$	0	$0.405 * Y/0.285$	0
Total deadweight losses DWL_i	$0.34 * X/0.27$	$0.31 * X/0.285$	0	$0.31 * Y/0.285$	0
Total PRTCs $PRTC_i$	$tc1 * X/0.27$	$tc2 * X/0.285$	$tc3 * X$	$tc4 * Y/0.285$	$tc5 * Y$

Y = transfers from a targeted policy; Z = transfers to unintended beneficiaries because of lack of targeting. W = transfers to unintended beneficiaries because of transfer inefficiency. $T = Y + Z + W + DWL$.
1. Deadweight losses and transfer efficiency ratios of Table 2.2.
Source: OECD Secretariat.

The comparisons to date are rather stylised in nature. Box 2.6 presents a set of calculations where the targeting ratio has been derived from actual structural and income data. A similar exercise could be carried out with respect to any policy objective. For example in the case of an area payment aimed at landscape preservation, the targeting ratio would relate to the share of the total agricultural area where it is desired to preserve the particular landscape value and where, to do so, a policy intervention is needed.

Figure 2.5. **Income support: Comparison of costs by policy type: illustration with a targeting ratio of 0.5, given assumptions on key parameters**

Illustrative purpose only

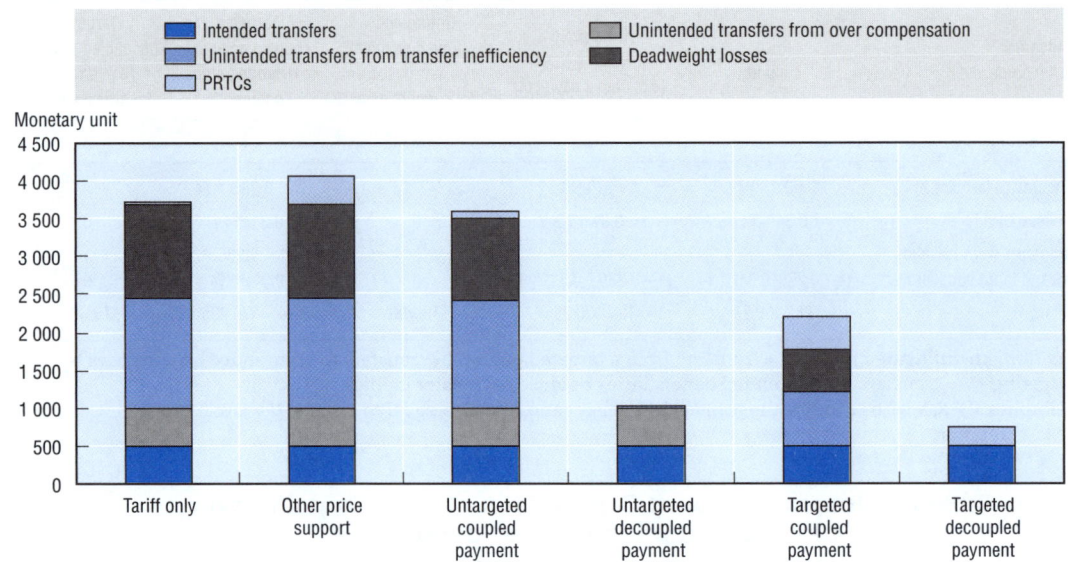

Source: Secretariat's calculations based on formulas in Table 2.6, unit PRTCs of the middle column of Table 2.2 and deadweight losses and transfer efficiency ratios of Table 2.3.

Box 2.6. **Numerical example of income targeting**

Income support is a widespread objective in OECD countries but most support is delivered in ways that are linked to production and goes to larger farms, often the richest ones (OECD, 2003b). Such support is therefore very inequitable and cost-inefficient in providing a safety-net to low-income households. Payments targeted to low-income households and triggered by income falling below a certain limit would be more efficient.[1,2] A targeted programme would allow savings as only those with an income lower than some defined objective level would receive support and would also improve income distribution and equity. Based on structural information, the amount of targeted transfers that would bring low incomes to the objective level can be estimated. It can then be compared to the amount that would achieve the same result for the lowest income group, using market price support. The ratio between the two estimates is the targeting ratio.

Data for average support, farm income and farm household income per farm or household, by quartiles based on gross sales, are used in this exercise.[3] In Table 2.7, the amount of additional transfers needed to bring the average household income of all four quartiles to the level of urban household income is estimated, first for a targeted measure that only compensates the farm household groups under the objective limit, second for a MPS measure that would bring the lowest quartile at parity level with urban households and would be distributed as current MPS. The top quartile that in the first case would not need additional transfers receives over 40% of all transfers in the second case and all transfers are four times what they would have been with targeting. In this illustrative example, the resulting degree of targeting of a measure that precisely targets low-income household groups is 25%. Total costs are presented in Table 2.8 and Figure 2.6.

THE IMPLEMENTATION COSTS OF AGRICULTURAL POLICIES – ISBN 978-92-64-03091-6 – © OECD 2007

Box 2.6. **Numerical example of income targeting** (*cont.*)

Table 2.7. **Estimation of additional support to reach income parity using quartiles based on gross sales**

Quartiles (Qi)	1	2	3	4	All farms	Total Sum (Qi * Ni)
	Average per farm ('000 won)					Billion won
Number of farms (Ni)	300 553	300 553	300 553	300 553	1 202 212	
Total output	9 859	14 913	21 211	32 519	19 710	
Direct payments	177	185	193	260	197	245
MPS	6 992	10 658	15 045	22 679	13 940	16 643
Farm income	1 099	5 146	8 353	14 934	8 048	8 876
Farm housed income (FHli)	5 021	13 508	20 921	38 542	20 223	23 441
Additional support to reach parity with urban households (UHI = 28 643 000 won in 2000)						
Targeted support (UHI – FHli)	23 622	15 135	7 722	n.a.	11 620	13 969
MPS as currently distributed	23 622	36 007	50 828	76 619	46 769	56 226
Distributional losses						42 257

n.a.: not applicable as FHI > UHI.
1. Additional MPS in Qi = Current MPS in Qi * Additional MPS in Q1/Current MPS in Q1 (average per farm).
Source: OECD structural data for Korea in 2000.

Table 2.8. **Application to income policy comparison**

For given outcomes (*e.g.* minimum income level)	Price support	Untargeted coupled payment	Untargeted decoupled payment	Targeted coupled payment	Targeted decoupled payment	Policy with no transfer
Transfers	208 245	197 284	56 226	49 015	13 969	0
Unintended transfers	123 472	122 157	42 257	19 851	0	0
– Within the sector (from lack of targeting)	42 257	42 257	42 257	0	0	0
– From transfer inefficiency	81 215	79 900	0	19 851	0	0
Deadweight losses	70 803	61 158	0	15 195	0	0
PRTCs	2 082	5 919	1 687	12 254	6 985	0

Source: Formulas from Table 2.6, parameters from Tables 2.2, 2.3 and 2.7.

1. USDA (2000) for example compares the cost of alternative safety-net programmes targeting farm households that fall below a certain income criteria with the cost of current programmes. It finds that, even when alternatives have a higher total cost, the distribution of payments is markedly different as they benefit poor households.
2. They could be agriculture-specific or managed through the general safety-net, in which case marginal PRTCs would be lower.
3. Korean data for 2000 are used for the purely illustrative purpose of comparing policies that aim to bring farm household income to parity with income of other households. The choice of data is quite arbitrary and does not reflect any stated objective of the Korean government.

Box 2.6. **Numerical example of income targeting** (cont.)

Figure 2.6. **Comparison of total costs by policy type: Graphical illustration of income policy**

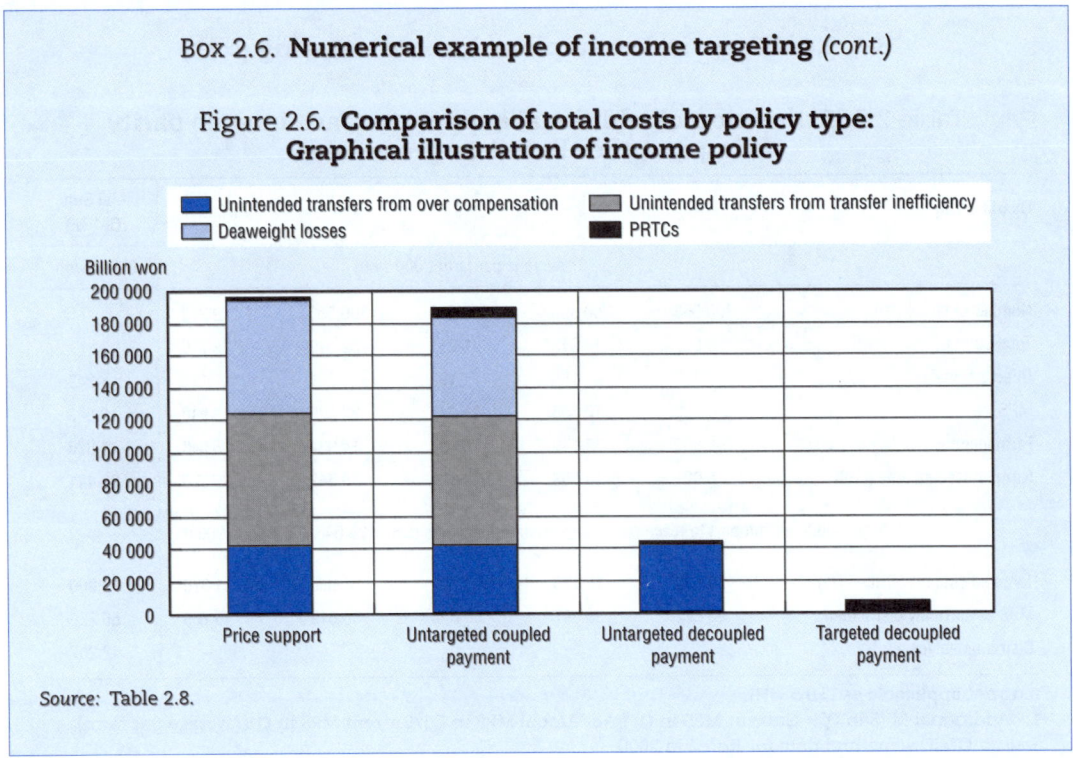

Source: Table 2.8.

Notes

1. In practice one usually refers implicitly to the Kaldor compensation criterion. In its simplified form this states that a policy change is commendable if it is potentially possible for the winners to compensate the losers. For small policy interventions that do not affect relative prices, a practical corollary is that a policy move is welfare enhancing if national income increases (Varian, 1992).

2. For the sake of simplification, no attempt is made to value outcomes in this analysis, which compares policies with the same outcome. Transfers are considered because the context is one of policy reform under financial constraints. In no case are transfers considered as a measurement of outcomes. They are a means to obtain the desired outcome. In the case of income policies, the level of transfers can be higher than the increase in farm income because of transfer inefficiency. The level of transfers is also generally different from the value of non-commodity output production and negative externalities, as there is no market where they could be equalised. Even with targeted and tailored policies, the transfer is usually not calculated as the value of the outcome, but rather the actual cost of achieving the desired result.

3. When valuing the different costs of a policy, care should be taken to avoid double-counting deadweight losses.

4. See Chapter 6, OECD (2003a).

5. While in conventional cost-benefit analysis, all welfare gains and losses are given the same weight, it is possible to use different weights to take account of distributional issues. For example, losses to poor households could be given a higher weight than losses to rich households to reflect equity concerns. The problem is to find the appropriate weights, as discussed in Chapter 15 of OECD (2006a).

6. A modelling approach developed as part of the cause and effect project carried out under the umbrella of the Joint Working Party on agriculture and the environment uses a welfare maximisation function to analyse policy choices in the case of agri-environmental measures. The structure of the model is described in Lankoski and Ollikainen (2003).

7. This means that the non-commodity output can be supplied independently from commodity production.

8. While the marginal deadweight losses vary along the supply curve, the unit deadweight losses considered here are an average over a given price equivalent range.

9. They could be expressed per hectare covered in the case of a land-based measure, with the same reasoning.

ISBN 978-92-64-03091-6
The Implementation Costs of Agricultural Policies
© OECD 2007

PART I

Chapter 3

Summary and Conclusions

In the general context of agricultural policy reform, this study suggests ways to reduce PRTCs and explores the role of policy-related transaction costs (PRTCs) in defining the most efficient option for achieving policy objectives. Although PRTCs have largely been neglected in agricultural policy design and analysis for a long time, they are attracting increasing interest with the development of more targeted policies.

PRTCs are defined as all costs arising from interactions between and within government agencies, private organisations and programme participants at all stages of policy implementation, starting with the initial gathering of information and the policy design, the selection of eligible farmers, the distribution of transfers, the monitoring and control stages, and ending with the final evaluation of the policy outcomes relative to objectives. For budgetary payments, implementation costs are sometimes strictly defined as the costs of delivering payments and monitoring farmers' eligibility and compliance. There are questions as to which costs can be considered as PRTCs. Technical assistance, for example, is sometimes excluded from PRTCs and considered as an outcome.

The characteristics of the policy, including the precision and clarity of its objectives or the nature of compliance, can influence PRTCs. Policy design can therefore be modified in order to reduce them. For a given policy, PRTCs also depend on the administrative structure and the regulatory environment, on structural factors such as the number, size and diversity of farms, and on access to information and co-ordination. Better knowledge of the factors affecting PRTCs would facilitate the estimation of PRTCs for future policies and therefore the design of best policies.

Everything else being equal, it will always be beneficial to try to reduce PRTCs, both in order to make better use of public funds, and to minimise one of the cost components of overall economic costs and benefits of a given programme. However, PRTCs must be incurred in pursuit of any government programme and are not "wasteful" per se. Some can even be considered as beneficial investment, e.g. capacity building or technical assistance. No government policy can be put in place without some PRTCs being incurred. It is, however, often possible to reduce PRTCs while maintaining benefits. Approaches to help reduce PRTCs include sharing experience, integrating policies and administrative networks, reducing the number of agencies, integrating government and private information systems, tendering contracts for specific tasks, and using information technologies. It is also recognised that making PRTCs transparent and monitoring them can enable them to be reduced. Finally, PRTCs for a given policy seem to decrease with time as experience accumulates.

There is wide recognition that PRTCs need to be taken into account when designing policies, together with all the other costs and benefits, and that they should be considered at an early stage of the policy design because they may modify the characteristics of the optimal solution. Taking PRTCs into account would help to identify those parameters that are important in any effort to reduce PRTCs. It could also help identify programmes that have become obsolete as their costs overcome their benefits.

While the importance of PRTCs is recognised, there are no examples of PRTCs being taken formally into account, together with other costs and benefits, to influence policy choice between large categories of policies. Few attempts have been made at estimating them and measurement is mostly made *ex post*, with varying degrees of reliability. In order to obtain more consistent and reliable estimates that could be used in policy comparison, much more systematic and accurate procedures would have to be used to measure PRTCs and evaluate policies.

The role of PRTCs in policy choices requires careful examination. PRTCs are only one component of the whole cost-benefit equation. Policy comparison requires all costs and benefits to be taken into account, including resource costs (deadweight losses, additional cost of de-linkage, *i.e.* the extra cost of producing a non-commodity output separately from commodity production – to be added to the transfer to producers needed to produce it jointly – and PRTCs), desired outcome, and other positive and negative externalities. In the context of sectoral policy reform where government want to pursue objectives at least financial cost, transfers generated by the policy should also be considered, in particular unintended transfers. PRTCs therefore cannot alone determine policy choice. No example was found in the literature of a full cost-benefit comparison of the type described here being carried out. When PRTCs are compared, it is often in unit terms, *e.g.* as a percentage of transfers, and there is no consideration of total PRTCs.

It is well beyond the purpose of this study to carry out complete cost-benefit analysis. The comparative analysis presented in Chapter 2 is schematic and purely illustrative. It focuses on some economic costs, assuming for the sake of simplification that all policies compared have the same desired outcome. Policy transfers, as well as resource costs (including PRTCs), are considered to be an important element in policy choice because the context is one of policy reform where governments want to achieve policy objectives under financial (budget) constraints. It is therefore important to consider what society is prepared to pay for obtaining specific outcomes. Welfare elements relate to economic efficiency, while transfers relate to distribution. Policy-makers may wish to consider these aspects separately. In some cases, however, the choice will be indeterminate. Policy-makers might then want to weight the two types of costs to reflect equity, feasibility and other social concerns, thus affecting policy choice. In this study, all elements are given a weight of one and different policy options are compared on the basis of the sum of their resource costs (including PRTCs) and transfers.

Assumptions are made regarding parameters and in particular the value of unit PRTCs. The median value of unit PRTCs is based on estimates found in the literature and case studies. Such estimates often focus on implementation costs while farmers' costs, information costs and evaluation costs are neglected. Farmers' costs are likely to be higher for conditional payments than for market price support. The costs of identifying targets are attributed to targeted policies. But in the case of broad-based policies with multiple objectives, the cost of identifying the targets (and the possible degree of jointness), is not taken into account, though it should be. In both cases, the costs of evaluating outcomes (for example the maintenance of biodiversity through the restoration of natural habitats) should be considered. Efforts have recently increased in this regard. For example, preservation of biodiversity might be a stated objective of a broad-based policy but its outcomes in terms of biodiversity are not necessarily evaluated. For a multiple-objective policy, evaluation should be carried out for all the outcomes to be verified. Another assumption regarding PRTCs is that unit PRTCs are fixed, whatever the amount of transfers.

The assumption that all options have the same outcome is very schematic. In reality, broad-based policies are more likely to miss their target than more specific policies. With a broad-based policy, more or less of the positive good than is demanded may be produced and there may be overcompensation or under-compensation. Moreover, the value of other positive and negative externalities may also change. It is therefore likely that specific outcomes cannot be achieved with broad-based policies without additional cross-compliance regulations, which would increase PRTCs, as well as expected benefits. In reality, efforts should be made to evaluate the outcome and changes in the value of positive and negative externalities under alternative policy options.

All other things being equal for a given outcome, the trade-off is clearly between the targeting ratio (i.e. the transfers to intended beneficiaries divided by the transfers to both intended and unintended beneficiaries) and the PRTCs. All the hypothetical examples developed in Chapter 2 show that the potential reduction in transfers as a result of targeting is a crucial parameter in policy choice. Although the PRTCs of targeted payments can be higher as a percentage of transfers than those of untargeted measures, total PRTCs are not necessarily higher. The parameters used in the hypothetical examples developed in this study help us define some limits, although they are not strictly speaking empirical in nature. They indicate that targeting, whether the policy is decoupled or not, is the least-cost option under a wide range of assumptions about key parameter values, especially when the targeting ratio is low. However, they may not reflect the wide diversity and complexity of situations found in OECD countries and thus the range of actual parameter values. One could envisage cases, where, for pursuing a policy objective, a targeted option does not have the lowest cost because of high PRTCs and/or a high targeting ratio. There may also be cases where the total cost of pursuing a policy objective is not lower with decoupled measures than with coupled measures due to high PRTCs and/or high costs of de-linkage. Moreover, the assumptions made to simplify the comparison should be kept in mind when interpreting results.

For example, an untargeted, decoupled policy with PRTCs of 1% of transfers is less costly than a targeted, decoupled policy that requires half the transfers, only if the PRTCs of the targeted policy are above 100%. This is an estimate rarely found in the literature, in particular for running costs. But if targeting only reduces transfers by 10%, the PRTCs of the targeted, decoupled measure should be lower than 12% of transfers for it to be preferred to a broad-based, decoupled option with PRTCs equal to 1% of transfers. When compared with a broad-based and coupled option with PRTCs of 1% of transfers, only if the targeted, decoupled option has PRTCs lower than 50% of transfers (base assumption on deadweight losses) would it be preferred. In other words, a broad-based, coupled policy with PRTCs of 1% would have lower total costs than a targeted policy if the targeted policy only reduces transfers by no more than 10% and if its PRTCs are over 50% of the value of transfers. In the case of income policies, the inclusion of transfer efficiency in the comparison reinforces the benefits of targeting as leakages from untargeted policies are particularly large, and confirm the benefits of decoupling. In the case of policies that aim to correct market failures when jointness exists, trade-offs between gains from decoupling and possible additional costs of de-linkage also need to be considered. This also means that the trade-off includes the transfers to producers needed for joint production of the public good on one side, and the total cost of separate production of the public good on the other side.

Unless the problem prompting government intervention is widespread and farm structure homogeneous, the conditions under which PRTCs could be so high as to offset the costs of distorting and over-compensating policies are quite specific, but care should be taken to ensure that they do not occur. The less widespread (more local) the problem is, the higher the gain from targeting. Conversely, gains from targeting may be low if there is strong jointness, the outcome sought is widespread (in the sense of relating to all production or all land) and targeting costs are high. The optimal option will therefore depend on the nature of the objectives and the characteristics of agricultural systems. Between policy options that are similar, the detail of implementation parameters may more often change the hierarchy. In any case, policy choice is currently being made in the absence of information on PRTCs, although they are measurable at a reasonable cost. Many issues, however, need to be further explored to improve policy comparison. Efforts should be made to estimate PRTCs in a more consistent and systematic way, in particular to explore the time dimension and the impact of institutional settings, but also to better evaluate costs and benefits, such as resource costs and quality and quantity of expected and unexpected outcomes.

PART I

References

Allen, D.W. (1991), "What are Transaction Costs?", *Research in Law and Economics*, Vol. 14, pp. 1-18.

Alston, J.M. and B.H. Hurd (1990), "Some Neglected Social Costs of Government Spending in Farm Programs", American Journal of Agricultural Economics, Vol. 72, pp. 149-156.

Arrow, K.J. (1969), "The Organisation of Economic Activity: Issues Pertinent to the Choice of Market *versus* Non-market Allocation," in *The Analysis of and Evaluation of Public Expenditure*, The PPB System 1, US Joint Economic Committee, 91st Congress, Washington DC, US Government printing office.

Ballard, C.L. and D. Fullerton (1992), "Distortionary Taxes and the Provision of Public Goods", *The Journal of Economic Perspectives*, Vol. 6, No. 3, Summer, p. 117-131.

Berriet-Solliec M., C. Déprès et D. Vollet (2003), "La multifonctionnalité de l'agriculture entre efficacité et équité : le cas des contrats territoriaux d'exploitation en Auvergne", *Économie rurale*, No. 273-274, January-April.

Browning, E.K. (1987), "On the Marginal Welfare Cost of Taxation", *American Economic Review*, Vol. 77, pp. 11-23.

Burgaz Moreno, F.J. (2003), *La administración del riesgo en la agricultura en el siglo XXI: la experiencas española,* presented at International Conference on Risk Management in Agriculture in the 21st Century, 25-26 September 2003, Santiago, Chile.

Carpentier, C.L., D.J. Bosch and S.S. Batie (1998), "Using Spatial Information to Reduce Costs of Controlling Agricultural Nonpoint Source Pollution", Agricultural and Resource Economics Review, Vol. 27, No. 1, April, pp. 72-84.

CBO (2004), *Administrative Costs of Private Accounts in Social Security*, The Congress of the United States, Congressional Budget Office, March.

Challen, R. (2000), Institutions, Transaction Costs and Environmental Policy: Institutional Reform for Water Resources, Edward Elgar Publishing, Inc., Massachusetts, USA.

Challen, R. (2001), *Non-government Approaches to the Provision of Non-Commodity Agricultural Outputs: A Transaction Cost Perspective*, paper presented at the OECD Workshop on "Multifunctionality: Applying the OECD Analytical Framework – Guiding Policy Design", Paris, 2-3 July.

CNASEA (2003), "L'application du règlement de développement rural en Europe : étude comparative", *Les Cahiers du CNASEA*, No. 3, April.

Coady, D.P. (2000), The Application of Social Cost-benefit Analysis to the Evaluation of PROGRESA, Final report, IFPRI, Washington, November.

Falconer, K. (1998), *The Transactions Costs of Countryside Stewardship Policies in 8 EU Member States*, TASK 3 report to the STEWPOL (FAIR1/CT95-0709) meeting, April, Durham.

Falconer, K. and M. Whitby (1999a), "The Invisible Costs of Scheme Implementation and Administration", in G. Van Huylenbroeck and M. Whitby, *Countryside Stewardship: Farmers and Markets*, Chapter 4, Elsevier Science Ltd.

Falconer, K. and M. Whitby (1999b), *Administrative Costs in Agricultural Policies: The Case of the English Environmentally Sensitive Areas*, University of Newcastle upon Tyne, Centre for Rural Economy, Research Report, June.

Falconer, K. (2000), "Farm-level Constraints on Agri-environmental Scheme Participation: A Transactional Perspective", *Journal of Rural Studies*, Vol. 16, pp. 379-394.

Falconer, K., P. Dupraz and M. Whitby (2001), "An Investigation of Policy Administrative Costs Using Panel Data for the English Environmentally Sensitive Areas", *Journal of Agricultural Economics*, Vol. 52, No. 1, pp. 83-103.

Falconer, K. and C. Saunders (2002), "Transaction Costs for SSSI and Policy Design", *Land Use Policy*, Vol. 19, Issue 2, pp. 157-166.

Feldstein, M. (1995), "The Effect of Marginal Tax Rates on Taxable Income: A Panel Study of the 1986 Tax Reform Act", *Journal of Political Economy*, Vol. 103, pp. 551-572.

Fraser, R.W. (2004), "On the Use of Targeting to Reduce Moral Hazard in Agri-environmental Schemes", *Journal of Agricultural Economics*, Vol. 55, No. 3, pp. 525-540.

Furubotn, E.G. and R. Richter (1998), *Institutions and Economic Theory*, The University of Michigan Press, USA.

GAO (1993), *Value-Added Tax: Administrative Costs Vary with Complexity and Number of Businesses*, United States General Accounting Office, Report to the Joint Committee on Taxation, United States Congress, Tax policy, GAO/GGD-93-78, May.

Hansson, I. and C. Stuart (1985), "Tax Revenue and Marginal Cost of Public Funds in Sweden", *Journal of Public Economics*, Vol. 27, No. 3, pp. 331-353.

Hazell, P.B.R. (1992), "The appropriate role of agricultural insurance in developing countries", *Journal of International Development*, Vol. 4, pp. 567-581.

IEEP (2001), *The Nature of Rural Development: Towards a Sustainable Integrated Rural Policy in Europe*, A ten-nation scoping study for WWF and the GB countryside agencies, Synthesis Report by David Baldock, Janet Dwyer, Philip Lowe, Jan-Erik Petersen and Neil Ward, January.

Johansson, P.O. (1991), *An Introduction to Modern Welfare Economics*, Cambridge University Press, Great Britain.

Ker, A.P. (2001), "Private Insurance Company Involvement in the US Crop Insurance Program", *Canadian Journal of Agricultural Economics*, Vol. 49, pp. 557-566.

Lankoski, J. and M. Ollikainen (2003), "Agri-environmental Externalities: A Framework for Designing Targeted Policies", *European Review of Agricultural Economics*, Vol. 30, No. 1, pp. 51-75.

Latacz-Lohmann, U. and C.P.C.M. Van der Hamsvoort (1998), "Auctions as a Means of Creating a Market for Public Goods from Agriculture", *Journal of Agricultural Economics*, Vol. 49, No. 3, September, pp. 334-345.

Mann, S. (2002), "The Concept of Administrative Elasticity", *International Journal of Public Administration*, Vol. 25, No. 8, pp. 1007-1019.

Mann, S. (2001), "Zur Effizienz der deutschen Agrarverwaltung", *Agrarwirtschaft*, Vol. 50, No. 5, pp. 302-307.

Mann, S. (2000), "Transaktionskosten der landwirtschaftlichen Investitionsforderung – ein komparativer Ansatz", *Agrarwirtschaft*, Vol. 49, No. 7, pp. 259-269.

Matthews, R.C.O. (1986), "The Economics of Institutions and the Source of Growth", *Economics Journal*, Vol. 96, pp. 903-18.

Mayshar, J. (1991), "Optimal Taxation with Costly Administration", *Scandinavian Journal of Economics*, Vol. 73, pp. 75-88.

McCann, L. and K.W. Easter (1999), "Transaction Costs of Policies to Reduce Agricultural Phosphorous Pollution in the Minnesota River", *Land Economics*, Vol. 75, No. 3, August, pp. 402-414.

McCann, L. and K.W. Easter (2000), "Public Sector Transaction Costs in NRCS Programmes", Journal of Agricultural and Applied Economics, Vol. 32, No. 3, December, pp. 555-563.

Millock, K., D. Sunding and D. Zilberman (2001), "Regulating Pollution with Endogenous Monitoring", *Journal of Environmental Economics and Management*, Vol. 44, pp. 221-241.

Miranda, M.J. (1991), "Area-Yield Crop Insurance Reconsidered", *American Journal of Agricultural Economics*, May.

Myers, R.J. (1992), "Can the Government Operate Programs Efficiently and Inexpensively?", *Contingencies*, March/April, pp. 15-17.

OECD (1998), *Agriculture in a Changing World: Which Policies for Tomorrow?* Press Communiqué, Meeting of the OECD Committee for Agriculture at Ministerial level [SG/COM/NEWS(98)22], 5-6 March, Paris.

OECD (2000a), *Valuing Rural Amenities*, OECD, Paris.

OECD (2000b), Income Risk Management in Agriculture, OECD, Paris.

OECD (2001a), Multifunctionality: Towards an Analytical Framework, OECD, Paris.

OECD (2001b), *Transaction Costs and Multifunctionality: Main Issues,* background paper to the OECD Workshop on "Multifunctionality: Applying the OECD Analytical Framework – Guiding Policy Design", Paris, 2-3 July.

OECD (2001c), Market Effects of Crop Support Measures, OECD, Paris.

OECD (2001d), Improving the Environmental Performance of Agriculture: Policy Options and Market Approaches, OECD, Paris.

OECD (2001e), Tax and the Economy: A Comparative Assessment of OECD Countries, OECD Tax Policy Studies, Paris.

OECD (2001f), Decoupling: A Conceptual Overview, OECD, Paris.

OECD (2002), The Incidence and Transfer Efficiency of Farm Support Measures, AGR/CA/APM(2002)24/FINAL.

OECD (2003a), Multifunctionality: The Policy Implications, OECD, Paris.

OECD (2003b), Farm Household Income: Issues and Policy Responses, OECD, Paris.

OECD (2005a), Workshop on Policy-related Transaction Costs, 20-21 January, Paris (*www.oecd.org/agr/meet/prtc*).

OECD (2005b), Multifunctionality in Agriculture: What Role for Private Initiatives?, OECD, Paris.

OECD (2005c), Agricultural Policies in OECD Countries: Monitoring and Evaluation 2005, OECD, Paris.

OECD (2006a), *Cost-benefit Analysis and the Environment: Recent Developments*, by David Pearce, Giles Atkinson and Susana Mourato, OECD, Paris.

OECD (2006b), "Financing Agricultural Policies with Particular Reference to Public Good Provision and Multifunctionality", AGR/CA/APM(2005)19/FINAL.

Pigou, A. C. (1947), *A Study in Public Finance*, London, Macmillan, Third Edition.

Polman, N.B.P. (2002), Institutional Economics Analysis of Contractual Arrangements; Managing Wildlife and Landscape on Dutch Farms, PhD Thesis, Wageningen University.

SAI (2000), *Administration of Arable Payments in the Netherlands, Sweden and England*, A report by the Netherlands Court of Audit, the Swedish National Audit Office and the United Kingdom National Audit Office, April.

Salhofer, K. (1996), "Efficient Income Redistribution for a Small Country Using Optimal Combined Instruments", *Agricultural Economics*, Vol. 13, pp. 191-199.

Sandford, C.T. and M. Goodwin (1986), Administrative and Compliance Issues Unique to VAT: Lessons from Two Periods of British Experience, The World Bank DP DRD-192.

Sandford, C.T. and O. Morrissey (1985), *The Irish Wealth Tax: A Case Study in Economics and Politics,* Paper 123, The Economic and Social Research Institute, Dublin.

Skees, J.R. (2000), "Agricultural Insurance Programmes: Challenges and Lessons Learned", in OECD (2000b).

Stavins, R.N. (1995), "Transaction Costs and Markets for Pollution Control", *Resources*, Spring, pp. 9-10, 18-20.

Stuart, C. (1984), "Welfare Costs Per Dollar of Additional Tax Revenue in the United States", *American Economic Review,* Vol. 74, pp. 352-362.

Thompson, D.B. (1999), "Beyond Benefit-cost Analysis: Institutional Transaction Costs and Regulation of Water Quality", Natural Resources Journal, Vol. 39, No. 3, Summer, pp. 517-541.

UNDP (2000), *Aid Transaction Costs in Vietnam*, United Nations Development Programme, Department for International Development, Final Report, December.

Varian, H.R. (1992), *Microeconomic Analysis*, W.W. Norton and Company, Third Edition.

Vatn, A. (2001), *Transaction Costs and Multifunctionality*, paper presented at the OECD Workshop on "Multifunctionality: Applying the OECD Analytical Framework – Guiding Policy Design", Paris, 2-3 July.

Vatn, A., V. Kvakkestad and P.K. Rørstad (2002), *Policies for Multifunctional Agriculture: The Trade-off Between Transaction Costs and Precision*, Agricultural University of Norway, Department of Economics and Social Sciences, Report No. 23, ISSN 0802-9210.

Weaver, R.D. and Taeho Kim (2002), "Moral Hasard: The Achilles Heal of Insurance for Managing Ag and Food System Performance Risks?", presented at *Risk and Uncertainty in Environmental and Resource Economics*, 5-7 June, Wageningen University, The Netherlands.

Williamson, O.E. (1985), *The Economic Institutions of Capitalism*, Free Press, NY.

ISBN 978-92-64-03091-6
The Implementation Costs of Agricultural Policies
© OECD 2007

ANNEX I.1.

Main Findings from the Literature Review and Case Studies

Table I.1.1. **Summary of main studies estimating PRTCs**

Authors	Title	Objectives	Approach to estimate PRTCs	Main results
SAI (2000) Supreme Audit Institutions of the Netherlands, Sweden and the United Kingdom	Administration of arable area payments in the Netherlands, Sweden and England.	Compare the administration costs of implementing arable area payments in three countries of the EU, by region.	Estimation of total cost of administration and cost per application based on staff time and cost. Estimation of the average time to process a claim.	National estimates in Table I.1.2. Large difference between regions within a country due to both the share of simplified payments and internal management factors.
Falconer and Whitby (1999a) Falconer (ed), team of EU researchers	The invisible costs of scheme implementation and administration. EU research project (STEWPOL), Task 3: Transactions and administrative costs in countryside stewardship policies.	Estimate PRTCs for 40 agri-environmental measures in 8 EU member countries.	Database on PRTCs, interviews, estimation of time spent.	Estimates of PRTCs for various agri-environmental schemes in EU member states are presented in Tables I.1.4 and I.1.5. Estimations of PRTCs for area and livestock payments are also reported (Table I.1.3).
Falconer and Whitby (1999b) Falconer, Dupraz and Whitby (2001)	Administrative costs in agricultural policies: the case of the English Environmentally Sensitive Areas. An investigation of policy administrative costs using panel data for the English Environmentally Sensitive Areas.	Identify theoretically and empirically the determinants of PRTCs in agri-environmental management agreements.	From the UK National Audit Office.	The number of agreements, the area under agreement and the number of years since designation are significant explanatory variables of PRTCs.
Falconer (2000)	Farm-level constraints on agri-environmental scheme participation: a transactional perspective.	Estimate PRTCs incurred by farmers participating in EU agri-environmental schemes.	From STEWPOL project.	A more integrated approach to agri-environmental schemes, including a clarification of objectives, would contribute to a reduction in farmers' PRTCs.
Falconer and Saunders (2002)	Transaction costs for SSSI (Sites of Special Scientific Interest) and policy design.	Compare the PRTCs of individually negotiated and standard management agreements.	Observation of actual PRTC for a sample of case-study management agreements.	Standard management agreements have higher compensation and transaction costs, even though their negotiation costs are lower.
Carpentier, Bosch and Batie (1998)	Using spatial information to reduce costs of controlling agricultural non-point source (NPS) pollution.	Theoretical model including NPS control costs; applied to policies to reduce nitrogen runoff from dairies in Lower Susquehanna watershed.	Number of hours required for tasks and type of professionals performing tasks multiplied by hourly wages by type of professional.	Targeting based on spatial information reduces both compliance and transaction costs because fewer farms need to be contracted and therefore monitored.
Thompson (1999)	Beyond benefit cost analysis: Institutional transaction costs and regulation of water quality.	Compare a non-tradeable emission limit permit (US) and an effluent charges policy (Germany) to improve water quality (textile mills).	Rough estimation of differences in compliance, enactment, implementation, detection and prosecution costs.	PRTCs do not change policy choice and effluent charges are still preferable to non-tradable effluent limit when PRTCs are taken into account.
McCann and Easter (1999)	Transaction costs of policies to reduce agricultural phosphorous pollution in the Minnesota river.	Determine whether PRTCs help explain existing policy choice; Identify factors affecting PRTCs in the case of environmental policies.	Labour cost = staff time * average salary for the different categories of staff (farmer time not included). Staff interviews to identify time requirements for the different steps, for 4 proposed programmes.	Transaction costs vary widely between policies. They reinforce the efficiency advantages of an input tax relative to a standard practice.

Table I.1.1. **Summary of main studies estimating PRTCs** (cont.)

Authors	Title	Objectives	Approach to estimate PRTCs	Main results
McCann and Easter (2000)	Public sector transaction costs in NRCS programmes.	Examine the magnitude of PRTCs associated with NRCS programmes in the US; identify the determinants of PRTCs (regression).	Survey of 6007 National Resource Inventory points containing information on conservation practices, public implementation costs and private and public conservation costs. Administrative costs estimated as time * average salary.	Estimated PRTCs in Table I.1.6.
Mann (2000)	Transaktionskosten der landwirtschaftlichen Investitionsforderung – ein komparativer Ansatz.	Compare PRTCs for investment grants and tax rebates in three regions of Austria, Germany and Switzerland.	Public budget and organisation charts on the one hand; indirect measurement techniques (compensation payments of accompanying organisations) on the other hand.	Estimated PRTCs are presented in Table I.1.7.
Mann (2001)	Zur Effizienz der deutschen Agrarverwaltung.	Identify institutional factors explaining PRTCs of all agencies involved in policy implementation in 131 German regions.	Idem.	Estimated PRTCs are presented in Table I.1.8. The two main factors increasing PRTCs are the multiplication of local agencies and the number of administrative layers.
Mann (2002)	The concept of administrative elasticity.	PRTC of export subsidies in Germany, related to expenditures.	Idem + number of staff.	Estimated PRTCs are presented in Table I.1.9.
Ker (2001)	Private Insurance company involvement in the United States crop insurance programme.	Compares the administrative costs of private *versus* public delivery systems.	Provided by delivery agencies or private companies.	Estimated PRTCs are presented in Table I.1.10. Rent-seeking behaviour may increase the administrative costs of privately-run insurance schemes, when those costs are subsidised by the government.
Skees (2000) Burgaz (2003)	Agricultural insurance programmes: Challenges and lessons learned.	Compares the administrative costs of insurance programmes in various countries.	Provided by delivery agencies or private companies.	Estimated PRTCs are presented in Table I.1.11.
Vatn, Kvakkestad and Rørstad (2002) Agricultural University of Norway	Policies for multifunctional agriculture: the trade-off between transaction costs and precision.	Examine theoretically the trade-off between PRTCs and gains from targeting; Estimate and compare PRTCs for various Norwegian agricultural policy measures.	Interviews with public servants, wholesalers and farmers to determine labour cost (time * cost/hr) and general overhead, computer, information material and postage costs. Running costs only. Extrapolation of local "representative" costs.	Estimated PRTCs are presented in Table I.1.12. Measures attached to commodities are found to have lower PRTCs as a percentage of transfers because their asset specificity is low and transaction frequency relatively high. Highest PRTCs as a percentage of transfers are found with measures that are site specific and have low frequency.

Table I.1.2. **Estimated costs and efficiency
of administration of area payments
in the Netherlands, Sweden and England**

	Netherlands (1997)	Sweden (1996)	England (1997-98)
Number of claims	52 000	59 600	46 800
Number of claims per regional office	10 400	2 480	5 210
Cost per application (EUR)	168	200	381
Cost of administration as a percentage of total payments to farmers	6.8%	2.7%	1%
Estimated average time to process a claim (hours)	2.2	3.7	12.7
Estimated range of processing times between regions (hours)	2.0-2.4	2.0-6.9	9.0-18.0

Source: Table 2.1 in SAI (2000).

Table I.1.3. **PRTCs of agricultural commodity regimes in Germany,
United Kingdom and Sweden**

As % of total public scheme cost

Germany (1993)	
Arable area payments	4
Livestock	20
UK (1996)	
Arable area payments	0.8
Set-aside	3.4
All crops and set-aside	1.4
Beef payments	4.9
Sheep	2.5
Sweden (1997)	
Arable area payments	3
Livestock	4

Source: Table 3.4 in Falconer and Whitby (1999a).

Table I.1.4. **Estimation of the PRTCs of agri-environmental programmes in the EU**

	PRTCs in EUR per ha	PRTCs in EUR per participant	PRTCs as a % of compensation payment	PRTCs as % of total cost
Austria (1996/97)	20.5	217	9	n.c.
Ecologically valuable area scheme			8.5	7.9
Eco-point scheme			11.2	10.1
Belgium (1996)	59	389	63	n.c.
Organic aid			2.5	2.4
E. Flanders willows scheme			66.3	39.9
Flemish management agreements (land consolidation)			8.8	8.5
Wallonia 2078			29.7	41.7
Bocage l'Ardennais			–	100
France (1996)	76	1 522	87	
Grass premium (Prime à l'herbe)			1.4	1-3
Conversion to extensive grassland			97	50
Long-term land retirement			110	52
Protection of endangered species			574	85
Reduction in animal density			80	44
Conversion to organic farming			15	53
Reduction in nitrogen use			45	31
Land restoration			302	75
Germany (1994/95)	10	177	12	
MEKA (market relief and cultural landscape)			1.1	1.1
SchALVO (payment for groundwater protection)			18	15.3
FUL group I (environmentally friendly agriculture)			3.7	3.5
FUL group II			118	54.2
Greece (1996-97)	60	470	9	
Reduction in nitrates			6.8	6.4
Organic aid			11.9	10.6
Long-term set-aside			7.5	7.0
Italy	13	140	7	
Sweden (1995)	9	190	11	
Valuable nature and cultural env.			12.7	11.3
Biodiversity in grazing lands			11.2	10.1
Open landscape			12.2	10.9
UK (1996/97)	48	2 446	48	
Env. Sensitive Areas (ESAs)			24.6	19.8
Countryside Stewardship (CSS)			38.0	27.6
NSAs			23.7	19.2
Moorland scheme			50.0	78.5
Habitat scheme			364.7	33.3
Organic aid			43.3	30.2
Countryside access			353.3	77.9
All 2078 measures			30.0	23.1

n.c.: not computable.

Measures considered in this table are very different in terms of targeting and specificity, thus the differences in PRTCs across measures. In addition, differences in institutional settings and delivery systems also contribute to differences across countries. Very high PRTCs as a percentage of transfers for some measures may be explained by set-up, fixed costs for newly introduced measures that are limited in scope and did not channel large transfers. Average annual cost of all listed programmes in front of the country's name.

Source: Table 3.3 in Falconer and Whitby (1999a).

Table I.1.5. **PRTCs of organic aid schemes in the EU**

		Annual PRTCs as a % of total costs	Hectares	Number of agreements	Annual PRTCs in EUR per ha	Annual PRTCs in EU per agreement
Austria	1995	7.37	259 588	18 543	1.3	18.2
	1996	7.35	272 062	19 433	1.3	18.6
	1997	7.33	280 000	20 000	1.4	19.0
Belgium	1994	3.68	2 219	95	8.3	194.8
	1995	3.58	2 690	109	8.9	219.5
	1996	2.39	3 591	143	5.8	146.3
England	1994	91.13	0	0	0	0
	1995	42.55	4 673	101	51.6	2 389.3
	1996	30.21	7 875	170	30.7	1 242.0
France	1994	28.35	18 850	732	51.9	1 337.7
	1995	27.61	20 324	783	50.1	1 300.3
	1996	29.79	32 331	1 417	55.9	1 275.4
Greece	1997	10.60	4 000	837	119.8	572.5

Source: Table 4.5 in Falconer and Whitby (1999a).

Table I.1.6. **PRTCs in National Resource Conservation Service (NRCS) programmes in the United States**

In USD	Mean	Standard deviation	Median
A. PRTCs per acre	12.52	30.33	3.31
B. Abatement cost per acre	20.32	50.84	3.00
C. Total conservation costs per acre (A + B)	32.84	62.87	10.73
A/C. PRTCs as a % of total costs	38	n.a.	n.a.

n.a.: not available.
Source: Table 2 in McCann and Easter (2000).

Table I.1.7. **PRTCs of agricultural investment subsidies in three regions of Austria, Germany and Switzerland**

	Ostprignitz/Oberhavel, Germany	Vorarlberg, Austria	Graubünden, Switzerland
PRTCs as a % of transfers	52	23	13

Source: Tables 2, 3 and 4 in Mann (2000).

Table I.1.8. **Total PRTCs per hectare and per farm in regions of Germany**
In DM

Region	PRTCs per hectare	PRTCs per farm
Bühl	184	–
Landau-Isar (Lower Bavaria)	–	1 170
Magdeburg	–	28 647
Reichelsheim	181	–
Krumbach (Swabia)	44	1 129
Wenigerode	–	19 003
Wittenburg	44	–

Source: Tables 1 and 2 in Mann (2001).

Table I.1.9. **PRTCs of export subsidies administration in Germany**

	Export subsidies	PRTCs	PRTCs as a % of transfers
	Million DM	Million DM	%
1991	5 127	22.6	0.44
1992	3 652	23.3	0.64
1993	3 425	24.7	0.72
1994	2 422	24.4	1.01
1995	2 210	24.9	1.13
1996	1 741	28.7	1.65
1997	1 687	28.8	1.71
1998	1 229	30.1	2.45
1999	1 431	30.0	2.10
2000	1 289	29.1	2.26

Source: Table 1 in Mann (2002).

Table I.1.10. **PRTCs of insurance programmes in North America**

United States	Total outlays (mil USD)	PRTCs as a % of total outlays
1990	673	40
1991	694	35
1992	624	39
1993	1 266	20
1994	302	97
1995	1 423	26
1996	1 398	35
1997	928	49
1998	1 459	29
1999	2 243	22
Canada		PRTCs as a % of premiums
Average 1990s		Around 15

Source: Ker (2001).

Table I.1.11. **PRTCs of insurance programmes in other countries**

	Period	Administrative costs as a % of premiums
Brazil[1]	1975-81	28
Costa Rica[1]	1970-89	54
Japan[1]	1947-77	117
	1985-89	357
Mexico[1]	1980-89	47
United States[1]	1980-89	55
United States[2]	1999	96
Spain[3]	1980-2002	18

1. *Source:* Hazell (1992).
2. *Source:* Skees (2000).
3. *Source:* Burgaz (2003).

Table I.1.12. **PRTCs for various programmes in Norway**

Policy instrument	Total transfer (million NOK)	PRTCs as a % of transfer
B1. Per hectare payment	3 267	1
B1. Livestock payment	2 088	2.3
A1. Output payments for milk	520	0.25
A1. Environmental tax on fertilisers	158	0.1
B1. Subsidy for reduced tillage	132	6.8
B3. Support for special landscape ventures	113	54
A2. Environmental tax on pesticides	53	1.1
B2. Per hectare payment to organic farming	19	18
B2. Conversion payment to organic farming	7	29
A2. Output payments for home-refined dairy products	1	12
B2. Payment for preserving cattle breeds	1	66

Source: Table 5.27 in Vatn *et al.* (2002).

Table I.1.13. **Allocation of PROCAMPO's PRTCs**

	PRTCs of ASERCA	PRTCs of CADERs	Total PRTCs	PRTCs as a % of payments	PRTCs per producer	PRTCs per hectare
	MXN mn	MXN mn	MXN mn	%	MXN	MXN
Tasks						
Design[1]	5	0	5	0.04	2	0
Evaluation	6	0	6	0.05	2	0
Identification of beneficiaries	0	48	48	0.37	17	4
Processing of applications	30	68	97	0.75	34	7
Actual payment	46	25	71	0.54	25	5
Eligibility/compliance	13	23	36	0.28	13	3
Monitoring[2]	115	0	115	0.88	40	8
Total	215	164	379	2.90	133	27

1. Does not include SAGARPA costs other than ASERCA's.
2. Includes management, computer operation, and organisation costs.
Source: Table 4.2 of Chapter 4.

THE IMPLEMENTATION COSTS OF AGRICULTURAL POLICIES – ISBN 978-92-64-03091-6 – © OECD 2007

Table I.1.14. **PRTCs of soil conservation programmes in the United States, 1983-2002**

Cost of technical assistance and administrative support	Initial year(s)	Conservation Reserve Program (CRP)	Wetland Reserve Program (WRP)	Environmental Quality Improvement Program (EQIP) and predecessors
		1986, 1997	1993	1995-96
Initial year(s), (million 1996 constant USD)	NRCS technical assistance	53.4	5.3	194.3
	FSA administrative support	62.4	n.a.	10.7
Succeeding year(s), (million 1996 constant USD)	NRCS technical assistance	353.2	85.5	1 476.5
	FSA administrative support	925.3	n.a.	168.5
In percentage of expenditure, initial year(s)	NRCS technical assistance	3	111	62
	FSA administrative support	4	n.a.	3
In percentage of expenditure, succeeding year(s)	NRCS technical assistance	1	9	37
	FSA administrative support	4	n.a.	4
Per acre enrolled, initial year(s), (1996 constant USD per acre)	NRCS technical assistance	23.21	106.93	n.a.
	FSA administrative support	27.11	n.a.	n.a.
Per acre enrolled, succeeding year(s) (1996 constant dollars per acre)	NRCS technical assistance	5.33	93.38	n.a.
	FSA administrative support	13.97	n.a.	n.a.

n.a.: not available.
Source: Table 6.5 of Chapter 6.

Table I.1.15. **PRTCs of direct payments in Canton Grisons**

Level	State	Canton	Boroughs	Farm	Total	Share of government PRTCs
Unit	CHF	CHF	CHF	CHF	CHF	%
Area payments	9 156	104 887	45 403	233 586	393 032	41
Payments for keeping grazing farm animals	8 597	95 893	41 582	213 878	359 950	41
Payments for keeping livestock under difficult conditions	8 666	95 895	41 581	215 430	361 572	40
Payments for farming on steep slopes	8 546	94 577	40 951	209 978	354 052	41
Payments for ecological compensation	9 012	113 789	44 610	232 779	400 190	42
Payments for extensive cereal and rapeseed cultivation	911	19 960	4 128	27 878	52 877	47
Payments for organic crop farming	5 951	71 610	19 382	477 074	574 017	17
Payments for animal housing systems	2 311	40 288	10 194	57 968	110 761	48
Payments for turning animals outdoors regularly	7 411	102 470	35 789	248 165	393 835	37
Total	60 560	739 369	283 622	1 916 736	3 000 287	36

Source: Table 5.8 of Chapter 5.

Table I.1.16. **PRTCs of direct payments in Canton Zurich**

Level	State	Canton	Boroughs	Farm	Total	Share of government PRTCs
Unit	CHF	CHF	CHF	CHF	CHF	%
Area payments	25 804	151 225	108 916	641 768	927 713	31
Payments for keeping grazing farm animals	10 130	57 569	41 313	247 566	356 578	31
Payments for keeping livestock under difficult conditions	3 643	19 923	13 958	88 171	125 695	30
Payments for farming on steep slopes	4 990	27 636	19 658	112 946	165 230	32
Payments for ecological compensation	26 188	153 419	111 305	671 223	962 135	30
Payments for extensive cereal and rapeseed cultivation	10 574	58 691	42 379	249 141	360 785	31
Payments for organic crop farming	2 140	9 305	6 585	184 832	202 862	9
Payments for animal housing systems	6 295	34 059	24 382	236 826	301 562	21
Payments for turning animals outdoors regularly	10 640	60 540	43 138	425 256	539 574	21
Total	100 404	572 368	411 635	2 857 729	3 942 134	28

Source: Table 5.11 of Chapter 5.

Table I.1.17. **Government PRTCs of direct payments in Cantons Grison and Zurich in relative terms**

Level	Total government PRTCs		PRTCs as a % of payments		PRTCs per farm		PRTCs per unit (ha or LSU)	
Unit	CHF		%		CHF per farm		CHF per ha or LSU	
Canton	Grisons	Zurich	Grisons	Zurich	Grisons	Zurich	Grisons	Zurich
Area payments	159 446	285 945	0.3	0.3	58	79	3	4
Payments for keeping grazing farm animals[1]	146 072	109 012	0.5	0.8	56	60	4	7
Payments for keeping livestock under difficult conditions	146 142	37 524	0.4	0.9	55	48	4	3
Payments for farming on steep slopes	144 074	52 284	1.0	2.2	56	55	4	10
Payments for ecological compensation	167 411	290 912	2.8	2.3	62	80	12	32
Payments for extensive cereal and rapeseed cultivation	24 999	111 644	7.8	4.4	92	70	31	17
Payments for organic crop farming	96 943	18 030	1.6	0.9	70	51	3	3
Payments for animal housing systems[1]	52 793	64 736	4.0	2.5	71	56	4	3
Payments for turning animals outdoors regularly[1]	145 670	114 318	1.9	1.5	62	58	3	3
Total	1 083 551	1 084 405	0.7	0.8	395	297	21	15

1. Unit is LSU (Livestock Unit) for those payments and ha (hectares for others).
Source: Tables 5.8, 5.9, 5.11 and 5.12 of Chapter 5.

Table I.1.18. **Evolution of the implementation costs
of the Common Agricultural Policies
in the Netherlands**

	PRTCS (million EUR)	EAGGF expenditures (million EUR)	PRTCs as a % of expenditures (%)
1990	33	3 020	1.1
1995	42	1 850	2.3
2000	56	1 400	4.0
2003	71	1 370	5.2
Market support	38	783	4.9
Direct payments	25	380	6.6
Rural development	8	205	3.9
2009	79	1 350	5.9
Market support	35	350	10.0
Direct payments	34	780	4.4
Rural development	10	220	4.5

The methodology used to estimate implementation costs and the grouping of CAP measures differ from the ones used in Table I.1.19.
Source: Presentation made by Tjeerd de Groot, from the Dutch Ministry of Agriculture, Nature and Food Quality, at the OECD Workshop on Policy-Related Transaction Costs.

Table I.1.19. **The administrative burden of agricultural policy
for Dutch farmers**

	Administrative burden as a % of expenditures(%)
Market price support	1.2
Income support	3.1
Product support	1.2
Project subsidies	22.7
Rural development (excluding projects)	2.8
National support	3.7
Total average	2.5

The methodology used to estimate administrative burden and the grouping of CAP measures differ from the ones used in Table I.1.18.
Source: Presentation made by Gerard de Vent, from the Dutch Ministry of Agriculture, Nature and Food Quality, at the OECD Workshop on Policy-Related Transaction Costs.

ANNEX I.2

The Marginal Costs of Taxation

The economic cost of raising one dollar of tax revenue is rarely one dollar. Pigou (1947) identified two costs of the tax system:

- the cost of administration and compliance; and

- the "... indirect damage (inflicted) on the taxpayers [...] over and above the loss they suffer in actual money payment."

This indirect damage results, at least in parts, from the fact that the tax system distorts relative prices. This welfare cost is often called "excess burden" or "marginal cost of public funds" in the literature. It is defined by Stuart (1984) as the per dollar surcharge that must be borne whenever the public sector alters the allocation or distribution of resources through fiscal measures.

The two types of costs have been examined in the literature. They are, however, difficult to estimate as briefly explained below. They depend on the characteristics of the taxation system such as the composition of tax by type (income, wealth or value-added tax, VAT), the marginal rates of taxation or the distributional impact of the tax structure, and on other economic factors such as the labour supply elasticity.

Many attempts have been made to estimate the marginal welfare cost of taxation, in particular for taxes on labour earnings. Estimates are very sensitive to assumptions about parameters, suggesting that marginal costs of taxation are difficult to estimate with any precision. Using a partial equilibrium framework, Browning (1987) estimates the marginal welfare cost of taxes on labour earnings in the United States to vary from under 10% to more than 300% of marginal tax revenue, depending on the marginal rate of taxation, the labour supply elasticity, the progressivity of the tax structure and whether earnings are unchanged or reduced by the tax change. Using a general equilibrium framework, Stuart (1984) finds estimates ranging from 7 to 133%, the higher estimates being for the higher rates of taxation. According to experts, the marginal social welfare cost of government spending is likely to be between 120 and 150% in the United States (Alston and Hurd, 1990). Estimates for Sweden also range widely, from 4 to over 100%, depending on assumptions on parameters (Hansson and Stuart, 1985). The estimate can be below zero under certain assumptions (Hansson and Stuart, 1985; Ballard and Fullerton, 1992). Estimations often focus on efficiency costs and ignore distributional effects of taxation.

In a comparative assessment of tax systems in OECD countries, OECD (2001e) considers tax administration and enforcement. Suggestions are made to raise the efficiency of tax collection and administration, and to reduce compliance and enforcement costs. Mayshar (1993) proposes a simple model of administration costs. His paper reports estimations of tax costs. The administration cost of raising VAT in England was 1% in the 80s (Sandford and Goodwin, 1986). The administration cost of wealth tax in Ireland ranged between 7 and 14% in the late 70s (Sandford and Morissey, 1985). Other estimates reported are even older. A Report by the United States General Accounting Office (GAO, 1993) estimated the administrative costs of various Value-Added Tax systems and found that they vary with complexity and number of businesses. Using United States panel data, Felstein (1995) finds that changes in marginal tax rates result in a very substantial response of taxable income (from 3 to 50% depending on marginal tax rates) due to changes in labour supply, tax reducing investments and compliance.

The marginal cost of taxation concept can be used to compare the distorting effects of two different tax options or, for a given tax, the marginal cost of raising funds can be compared to the benefits of spending the funds raised. In agricultural economics, the concept has been introduced in the context of policy comparison. Alston and Hurd (1990) and Salhofer (1995) discuss how the cost of public funds affects the efficiency ranking between price administration, quotas and deficiency payments.

The marginal cost of transfers from taxpayers is, in any case, less than the deadweight losses resulting from transfers from consumers of specific products because the former affects each sector in the same way and thus induces fewer distortions.

ANNEX I.3

Alternative Graphical Illustration

Figure I.3.1. **Graphical illustration of resource costs and unintended transfers**

Welfare maximisation and transfer minimisation

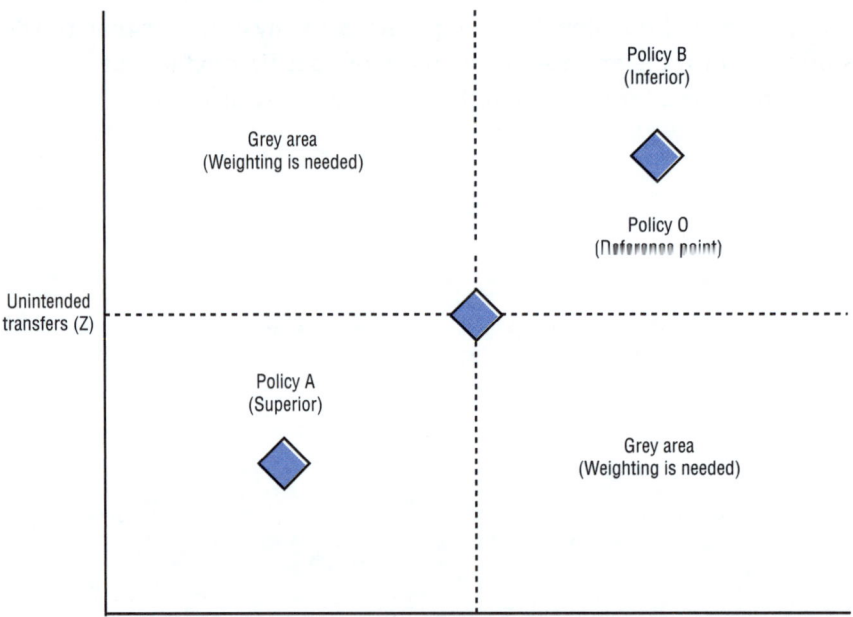

ANNEX I.4

Illustration with Alternative Parameters

The origin of the key parameters used in Figures I.4.1 and I.4.2 is explained in detail in Box 2.4. Ranges of plausible values for transactions costs are drawn from the literature review presented in Section 1.3 of Chapter 1, and estimates of the possible order of magnitude of deadweight losses associated with different policy instruments are drawn from OECD (2001c). For illustrative purposes transfers in the absence of targeting are assumed to be 1 000 monetary units (X = 1 000) and transfers of targeted measures are assumed to be 200, 500 or 800 monetary units (Y = 200, 500 or 800). This is defined as a targeting ratio of 0.2, 0.5 or 0.8. The additional cost of de-linkage is assumed to be 0%, 20% or 50%. A high additional cost of de-linkage would be linked to the existence of economies of scope which could imply strong jointness, meaning a strong positive relationship between commodity production and the production of the non-commodity output. A high targeting ratio would result from a widespread incidence of the non-commodity output. For the different policy instruments, the relative magnitude of the different cost elements that are retained for comparative purposes can be portrayed and possible trade-offs identified.

Figure I.4.1 first depicts resource costs on the horizontal axis and unintended transfers on the vertical axis, while in Figure I.4.2 the height of the columns represents the sum of the different costs that are included in the exercise. Assumptions for Figures I.4.1A and I.4.2A are that the targeting ratio is 50%, deadweight losses and PRTCs are maintained at their base level defined in Box 2.4 (Tables 2.3 and 2.4 respectively) and there are economies of scope leading to an additional cost of de-linkage equivalent to 20% of transfers. Assumptions for Figures I.4.1B and I.4.2B are that the targeting ratio is 80%, deadweight losses are reduced by 50% compared to the base assumptions, PRTCs increased by 50% compared to the base assumptions and there are economies of scope leading to an additional cost of de-linkage equivalent to 50% of transfers. Assumptions for Figures I.4.1C and I.4.2C are that the targeting ratio is 20%, base deadweight losses are increased by 50%, base PRTCs reduced by 50% and there are no economies of scope and thus no additional cost of de-linkage. In the case of perfect jointness, a coupled, targeted payment (for example a regional output payment) would have zero deadweight losses. For a broad-based, coupled policy such as price support, deadweight losses on the producer side would remain for unintended transfers and for all transfers on the demand side.

Figure I.4.2 shows that when the savings from targeting are large (assumptions A and C), the total cost of a targeted measure is likely to be lower than that of a broad-based measure. The choice between a coupled or a decoupled targeted option depends on the deadweight losses and the additional cost of de-linkage, given the assumptions made about PRTCs. Where there are significant additional costs of de-linkage or PRTCs and/or when deadweight losses are modest, a targeted, coupled payment costs less than a targeted decoupled measure. However, when the savings from targeting are smaller, the policy choice will depend on the level of deadweight losses, additional cost of de-linkage and PRTCs. When the targeting ratio is high (i.e. when a policy explicitly seeks to apply a common rate of support to almost all of the population, or to almost all land), the targeted option may not have the lowest cost (Figures I.4.1B and I.4.2B).

Figure I.4.1. **Market failure: comparison of resource costs** *versus* **unintended transfers by policy type based on different hypothetical combinations of key parameters**

Illustration purpose only

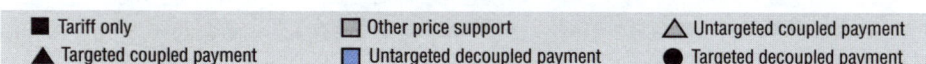

■ Tariff only   □ Other price support   △ Untargeted coupled payment
▲ Targeted coupled payment   ■ Untargeted decoupled payment   ● Targeted decoupled payment

A. Base assumptions

- Targeting ratio of 50%;
- Additional cost of de-linkage of 20% of transfers;
- Base estimates of deadweight losses (Table 2.3); and
- Base estimates of PRTCs (Table 2.2).

B. Strong jointness, widespread, lower deadweight losses and higher PRTCs:

- Targeting ratio of 80%;
- Additional cost of de-linkage of 50% of transfers;
- Base estimates of deadweight losses reduced by 50%; and
- Base estimates of PRTCs increased by 50%.

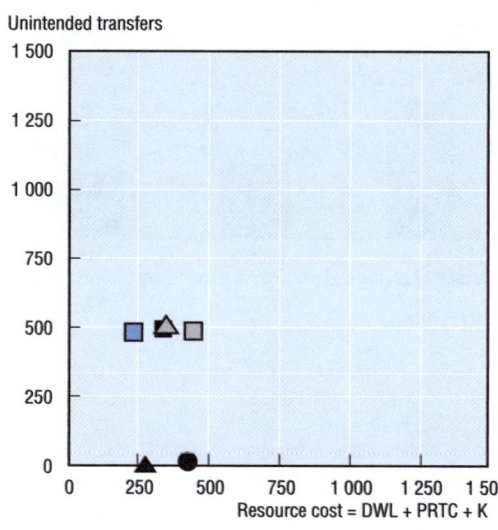

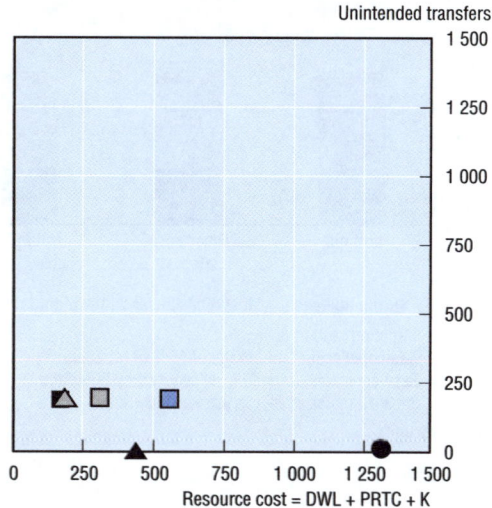

C. Weak jointness, not widespread, higher deadweight losses and higher PRTCs:

- Targeting ratio of 20%;
- No additional cost of de-linkage;
- Base estimates of deadweight losses increased by 50%; and
- Base estimates of PRTCs increased by 50%.

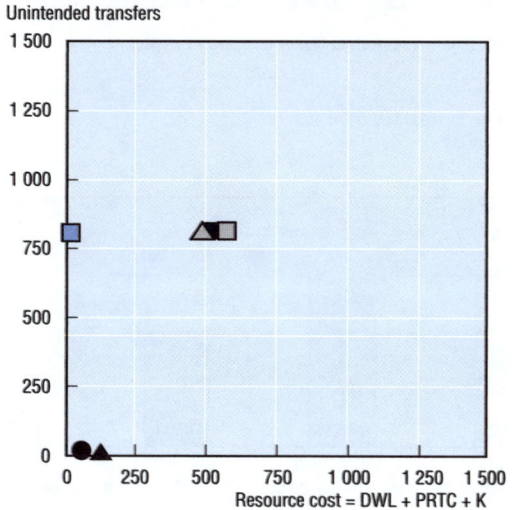

DWL: Deadweight losses; Z: Unintended transfers; K: Additional cost of de-linkage.

Source: Secretariat calculations based on formulas in Table 2.1.

Figure I.4.2. **Market failure: Comparison of costs by policy type, based on different hypothetical combinations of key parameters**

Illustration purpose only

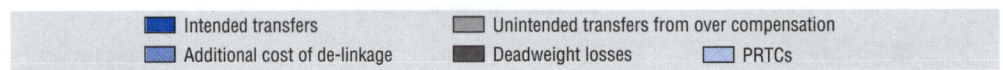

Intended transfers | Unintended transfers from over compensation
Additional cost of de-linkage | Deadweight losses | PRTCs

A. Base assumptions

- Targeting ratio of 50%;
- No additional cost of de-linkage of 20% of transfers;
- Base estimates of deadweight losses (Table 2.3); and
- Base estimates of PRTCs (Table 2.2).

Monetary unit

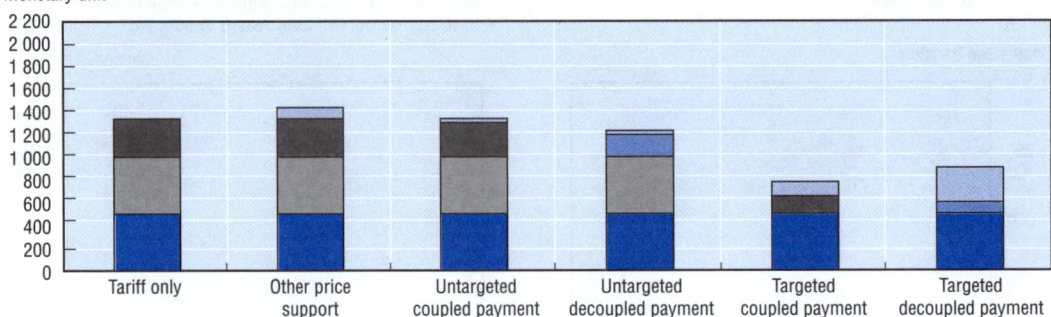

B. Strong jointness, widespread, lower deadweight losses and higher PRTCs:

- Targeting ratio of 80%;
- Additional cost of de-linkage of 50% of transfers;
- Base estimates of deadweight reduced by 50%; and
- Base estimates of PRTCs increased by 50%.

Monetary unit

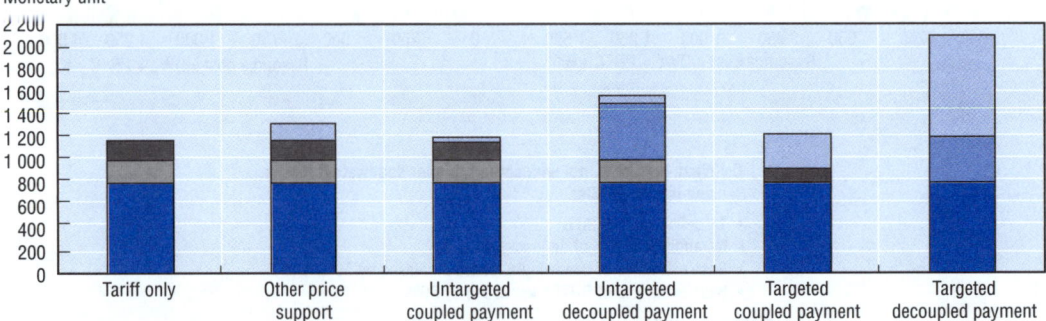

C. Weak jointness, not widespread, higher deadweight losses and lower PRTCs:

- Targeting ratio of 20%;
- No additional cost of de-linkage;
- Base estimates of deadweight losses increased by 50%; and
- Base estimates of PRTCs reduced by 50%.

Monetary unit

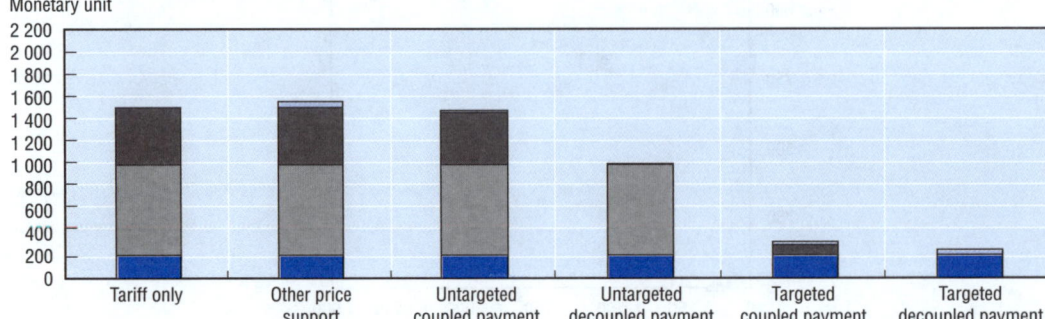

DWL: Deadweight losses; Z: Unintended transfers; K: Additional cost of de-linkage.

Source: Secretariat calculations based on formulas in Table 2.1.

Table I.4.1. **Market failure: %PRTCs and targeting ratios**
The choice between a targeted, decoupled policy (6) and an untargeted, coupled policy (1): Alternative values for DWL, based on different hypothetical combinations of key parameters

Maximum value of %PRTC (tc6) for the targeted option to be lower cost (%) (illustrative purpose only)

Targeting ratio (Y + K)/X											
PRTC of the untargeted, coupled policy as a % of transfers (tc1)	0.1	0.2	0.25	0.3	0.4	0.5	0.6	0.7	0.8	0.9	1
	Base assumption on the deadweight losses of the untargeted, coupled policy (dwl1 = 0.34)										
1	1 250	575	440	350	238	170	125	93	69	50	35
5	1 290	595	456	363	248	178	132	99	74	54	39
10	1 340	620	476	380	260	188	140	106	80	60	44
20	1 440	670	516	413	285	208	157	120	93	71	54
40	1 640	770	596	480	335	248	190	149	118	93	74
50	1 740	820	636	513	360	268	207	163	130	104	84
	Deadweight losses of the untargeted, coupled policy reduced by 50% (dwl1 = 0.34 * 0.5)										
1	1 080	490	372	293	195	136	97	69	48	31	18
5	1 120	510	388	307	205	144	103	74	53	36	22
10	1 170	535	408	323	218	154	112	81	59	41	27
20	1 270	585	448	357	243	174	128	96	71	52	37
40	1 470	685	528	423	293	214	162	124	96	74	57
50	1 570	735	568	457	318	234	178	139	109	86	67
	Deadweight losses of the untargeted, coupled policy increased by 50% (dwl1 = 0.34 * 1.5)										
1	1 420	660	508	407	280	204	153	117	90	69	52
5	1 460	680	524	420	290	212	160	123	95	73	56
10	1 510	705	544	437	303	222	168	130	101	79	61
20	1 610	755	584	470	328	242	185	144	114	90	71
40	1 810	855	664	537	378	282	218	173	139	112	91
50	1 910	905	704	570	403	302	235	187	151	123	101

$[tc6 * (Y + K) = X - Y - K + tc1 * X + dwl1 * x]$, thus tc6 in % = $100 * \{(1 + tc1/100 + dwl1)/[(Y + K)/X] - 1\}$.
X = transfers from an untargeted policy. Y = transfers from a targeted policy; K = additional cost of de-linkage.
Source: Based on formulas in Table 2.1.

Table I.4.2. **Market failure: %PRTCs and targeting ratios**
The choice between a targeted, decoupled policy (6) and an untargeted, decoupled policy (4) based on different hypothetical combinations of key parameters

Maximum value of %PRTC (tc6) for the targeted option to be lower cost (%) (illustrative purpose only)

Targeting ratio (Y + K6)/(X + K4)											
PRTC of the untargeted, decoupled policy as a % of transfers (tc4)	0.1	0.2	0.25	0.3	0.4	0.5	0.6	0.7	0.8	0.9	1
1	910	405	304	237	153	102	68	44	26	12	1
5	950	425	320	250	163	110	75	50	31	17	5
10	1 000	450	340	267	175	120	83	57	38	22	10
20	1 100	500	380	300	200	140	100	71	50	33	20
40	1 300	600	460	367	250	180	133	100	75	56	40
50	1 400	650	500	400	275	200	150	114	88	67	50

$[tc6 * (Y + K6) = (X + K4) - (Y + K6) + tc4 * (X + K4)]$, thus tc6 in % = $100 * \{(1 + tc4/100)/[(Y + K6)/(X + K4)] - 1\}$.
X = transfers from an untargeted policy. Y = transfers from a targeted policy; K = additional cost of de-linkage.
Source: Based on formulas in Table 2.1.

Table I.4.3. **Market failure: Illustration of multiple *versus*
single objective policies, using various assumptions
on %PRTCs and targeting ratios**

Maximum value of tc2 in % of transfers for two targeted policies
to be the best option (ilustrative purpose only)[1]

X1 (vertically) and X2 (horizontally) as a % of X0	10%	25%	50%	75%	90%
tc0 = 1%; tc1 = 10%					
10%	1 140	396	148	65	38
25%	975	330	115	43	
50%	700	220	60		
75%	425	110			
90%	260				
tc0 = 1%; tc1 = 20%					
10%	1 130	392	146	64	37
25%	950	320	110	40	
50%	650	200	50		
75%	350	80			
90%	170				
tc0 = 1%; tc1 = 50%					
10%	1 100	380	140	60	33
25%	875	290	95	30	
50%	500	140	20		
75%	125	-10^2			
90%	-100^2				

1. General formula: $X0 - X1 - X2 + (0.34 + tc0) * X0 = tc1 * X1 + tc2 * X2$ with $X1 + X2 < X0$
   $tc2 = [(1 + 0.34 + tc0) - (tc1 + 1) * X1/X0]/X2/X0 - 1$.
2. Negative numbers for tc2 indicate that if the PRTCs of the first targeted option are 50% of transfers, the savings in transfers from targeting must be over 25% for the two targeting options to be preferred to a coupled, broad-based policy.

Source: Secretariat's estimates.

PART II

Case Studies

ISBN 978-92-64-03091-6
The Implementation Costs of Agricultural Policies
© OECD 2007

PART II

Chapter 4

A Case Study of the Policy-related Transaction Costs of PROCAMPO Payments in Mexico

Executive Summary

Under the PROCAMPO programme of Mexico, eligible farmers receive payments based on the area planted during an historical base period (1991-93) on the condition that the land is used for legal agricultural or livestock production, or within an environmental programme. In 2002, PROCAMPO granted payments to over 2.7 million farmers for an area corresponding to 13.9 million hectares (i.e. 58% of the total agricultural area).

The PROCAMPO programme is administered by ASERCA, a decentralised body of the Secretariat for Agriculture, Livestock, Rural Development, Fisheries and Food (SAGARPA). ASERCA distributes and processes application forms, checks the eligibility of applicants, proceeds with the payment and maintains the database containing information on registered farmers, land use and payment levels. ASERCA also uses a Geographical Information System to monitor eligibility, check compliance and evaluate the environmental impact of the programme.

Local agencies of SAGARPA (CADERs) implement PROCAMPO and other programmes at the municipal level. They distribute information on programmes, announce the payment rates, help farmers fill in applications, check eligibility, collect application forms and send them to ASERCA regional offices. At the end of the procedure, they inform farmers of the amount they will receive and, in some cases, actually give them the cheque.

PROCAMPO uses different means of payment: cheques that can be cashed by bearers (called PROCAMPO cheques), transfers to deposit accounts, and transfers to withdrawal electronic cards. PROCAMPO cheques are used because a large number of farmers do not have a bank account. ASERCA has been trying to encourage the use of withdrawal electronic cards, which are the least costly means of payment, by subsidising them.

ASERCA and CADERs' policy-related transction costs (PRTCs) were estimated using publicly available information, such as budgets, the number of staff, wages, working time, and organisation charts. A number of assumptions had to be made regarding the allocation of costs by programme and task, and on the value of labour costs.

Total PROCAMPO PRTCs were estimated at MXN 379 million (USD 35 million), or less than 3% of the total value of payments. To give a more precise picture of their relative size, PRTCs are also related to the number of producers receiving payments and hectares covered. Average estimated PRTCs per producer are MXN 133 (USD 12.4) for an average payment of MXN 4 592 (USD 427). The average payment per ha is MXN 947 (USD 88) for an estimated PRTC of MXN 27 (USD 2.5).

The PRTCs of PROCAMPO are relatively modest, given the large number of farmers receiving payments, because transactions are relatively standard and most tasks are computerized. The highest PRTC comes from monitoring of the system, i.e. the administration and coordination of the whole system by the central ASERCA office. The second highest cost comes from the processing of applications, two-third of which occurs at the local level. Identification of the beneficiary has a relatively low cost as all PROCAMPO producers are registered in a database. This database is used to check eligibility, also at a relatively low cost, through the monitoring of successive applications and information that have already been registered in it. In addition, it is expected that the cost of proceeding with payments will decrease as more producers cash their payment by using an electronic card and because since 2005 programme participants are no longer required to register a claim for each cycle. While further efforts to reduce PRTCs (without adversely affecting the results) should be pursued, it should also be kept in mind that PRTCs are only one element to be considered when looking at the cost-efficiency of a policy or comparing policy options. Ideally, all costs and benefits should be taken into account in policy evaluation.

4.1. Background

Chapter 4 contains a case study on PROCAMPO payments in Mexico. It first provides a brief overview of the programme (Section 4.2). It then examines the implementation system and institutions, payment conditions and means of payments (Sections 4.3 to 4.5). It outlines the use of information technologies in the system and their role in cost savings (Section 4.6). Estimations of the costs of implementing PROCAMPO payments are presented in Section 4.7 and some concluding remarks on their size are made in Section 4.8.

4.2. Brief overview of the programme

The PROCAMPO[1] programme disburses payments to eligible farmers based on the area planted during an historical base period (1991-93) on condition that farms use their land for legal agricultural or livestock production, or for an environmental programme. The programme was set up in 1993/94 for a period of 15 years. It originally compensated producers for the elimination of guaranteed prices on support crops managed by the State-owned marketing agency, CONASUPO,[2] with the objective of allowing farmers to respond to market signals in a context of increasing trade openness, while providing a certain level of income. The programme is set-up on the basis of two crop cycles a year (winter/fall and spring/autumn) with farmers receiving a flat rate payment per eligible hectare.

In 2002, PROCAMPO granted payments to over 2.7 million farmers for an area of 13.9 million hectares (i.e. 58% of the total agricultural area). The spring/summer cycle concerns 2.3 million producers and 10.8 million hectares; the autumn/winter cycle 450 000 producers and 3.1 million hectares. The payment rate was MXN 873 (USD 91) per hectare in spring/summer 2002 and the annual total budgetary cost was MXN 12 420 million (USD 1 292 million), compared to MXN 8 665 million (USD 902 million) for ALIANZA programmes[3] and MXN 2 723 million (USD 284 million) for marketing payments per tonne of maize, wheat, sorghum, rice and other crops (OECD, 2003a). Payments per head of cattle (PROGAN) were introduced in 2003 on the same basis as PROCAMPO payments.

4.3. Implementation system and institutions

The PROCAMPO programme is administered by **ASERCA** (Support Services for Agricultural Marketing Agency) on behalf of the central administration. ASERCA is a decentralised body of the Secretariat for Agriculture, Livestock, Rural Development, Fisheries, and Food (**SAGARPA**). It was created to enhance the commercialisation of some crops (maize, wheat, rice and oilseeds) as a way to help producers to benefit from trade liberalisation and the opening of international markets. There is a central office with around 360 employees and 9 regional offices (down from 16 initially) with around 40 employees per office on average.[4]

The structure of the Central Office is shown in Diagram of Figure 4.1. The structure of regional offices depends on the region's characteristics, but typically a regional office would have a Director, a Deputy-Director and three departments:

- a department for information technologies (computer systems), which provides computer support and maintenance; controls all incoming and outgoing documents; receives claims and issue cheques;

- a department for the control and evaluation of programmes; and

- an administrative department which manages personnel and equipment.

ASERCA distributes and processes application forms, checks the eligibility of applicants and proceeds with the payment. The ASERCA database contains information on eligible farmers, land use, payment levels, etc.

ASERCA also manages the implementation of marketing payments for crops and, since 2003, PROGAN payments, and it is planned that it will administer all *Alianza* programmes in the future (SAGARPA will continue to be responsible for PROCAMPO development) as well as centralise all information.

The Directorate General (DG) for Information Services of ASERCA is in charge of designing questionnaires that gather information on producers. It is also in charge of "SIGA" (*Sistema de Información Geográfica ASERCA*), a system of geographical information based on satellite images used to check consistency between PROCAMPO claims and payments, and land use.

Local agencies of **SAGARPA**, the **CADERs** (*Centros de Apoyo al Desarrollo Rural*), implement PROCAMPO and other programmes at the municipal level. They are the smallest interface between the government and producers. There are 715 CADERs in Mexico with an average of four employees. They distribute information on programmes, announce the amount of payments, help farmers fill in applications, check eligibility, and collect application forms and send them to ASERCA regional offices. At the end of the procedure, they tell farmers how much they will receive and, in some cases, actually give them the cheque.

Traditionally, farmers have to register with PROCAMPO at the beginning of the season, fill in a form for each cycle and submit it to the CADER and, in some cases, pick up their cheque at the CADER. Recent developments have been implemented to reduce the number of transactions for farmers and CADERs, as explained in the next sub-section.

4.4. Payment conditions

Producers receive payments for the agricultural cycle for which they apply and they have to make a claim for each cycle. There are three ways in which PROCAMPO payments are delivered to farmers (ASERCA, 2002a):

- **Traditional PROCAMPO** applies to farms of over 5 hectares. The call for applications and the level of payment are announced by local offices after planting. Four weeks after sowing, the producer is requested to complete and send the application form,[5] which is verified by both the local offices and ASERCA. Approximately four weeks later, the local office (CADER) announces the availability of the cheque.

Figure 4.1. **Flow chart of central ASERCA**

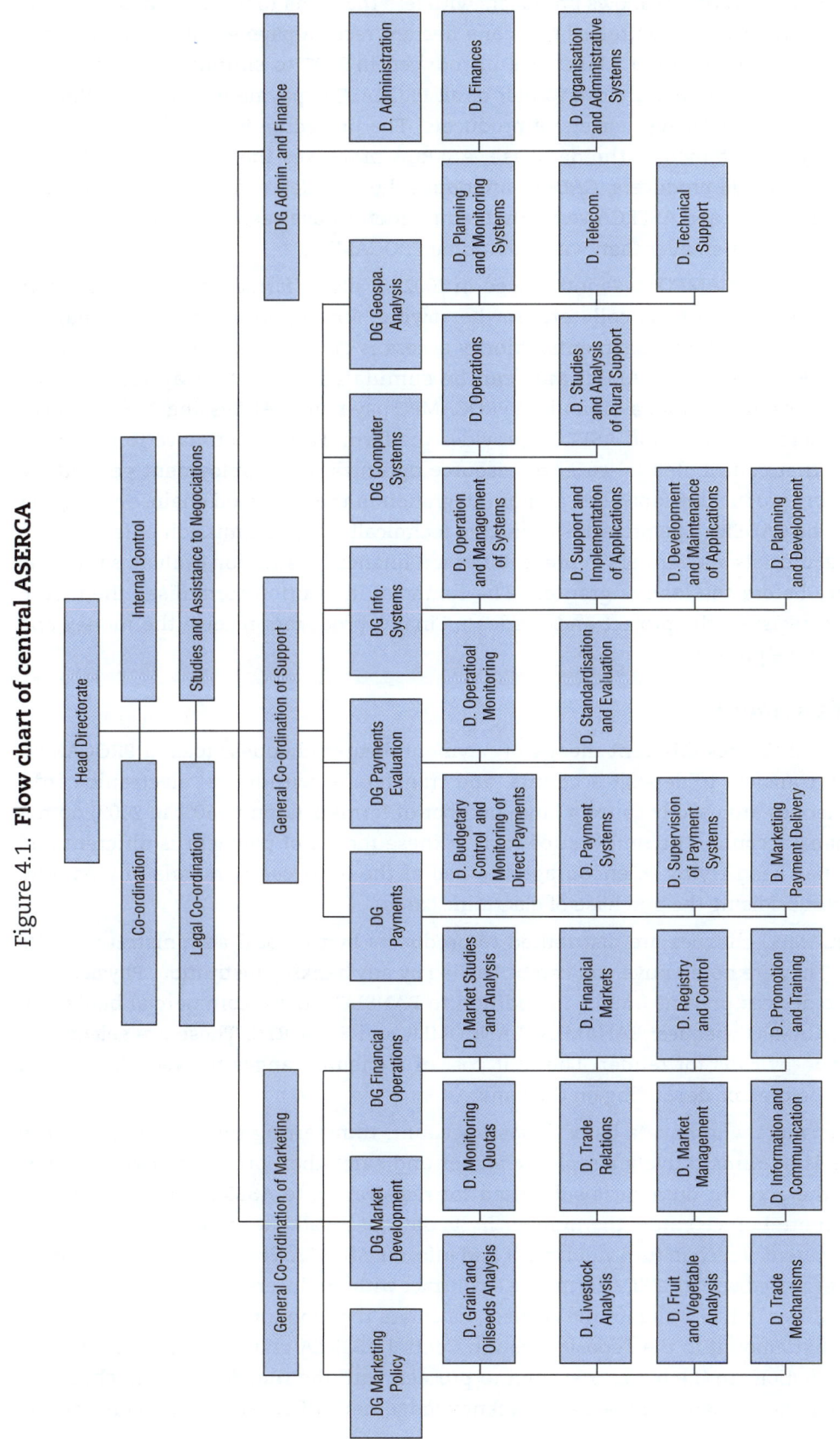

Source: ASERCA, *Estructura dictaminada,* 1 January 2002, Directorate General of Information Systems.

- **Anticipated PROCAMPO** allows producers with less than 5 ha to receive payment before sowing starts. Farmers with less than one hectare receive payment equivalent to one hectare. Anticipated PROCAMPO was introduced in 2001 to simplify administrative tasks. It now accounts for close to half of all PROCAMPO payments, over one-third of land, and close to three-quarters of producers. The procedure is as follows: based on information contained in the database, ASERCA processes the form and writes the cheques for small producers. CADERs announce the cheque's availability to producers, and both CADERs and ASERCA verify that requirements have been fulfilled. The number of transactions is smaller than with traditional PROCAMPO.

- **Capitalised PROCAMPO**[6] was introduced in 2002. It allows farmers to use their future PROCAMPO payments as collateral for borrowing. Small producers (up to 5 ha) are exempted from interest payments. Priority access is given to women and indigenous groups. As a rule, PROCAMPO aids can be cumulated with other agricultural aid programmes. To apply for a capitalised PROCAMPO payment, CADERs and ASERCA check the farmers' eligibility and ASERCA provides the form. Farms then have to propose a project that contributes to water or resource optimization, employment generation, increasing production capacities and/or integration in the agro-food chain, and register it with the CADERs. The project is evaluated technically by a commission of federal and regional officials and farmers' representatives. A financial institution evaluates financial viability before the loan is granted. The financial institution formalises the credit, ASERCA registers the project and the loan. Finally, producers receive the money and carry out the project.[7]

4.5. Means of payment

PROCAMPO uses different means of payment: bearer cheques[8] (called PROCAMPO cheques), transfers to deposit accounts, and transfers to withdrawal electronic cards. Bearer cheques are used because a large number of farmers (around 50% in 2003) do not have a bank account. The unit cost of each of these means of payment is different and ASERCA has been trying to encourage the use of the least costly means of payment, including subsidising the purchase of electronic cards.

PROCAMPO cheques are distributed to producers by the local administrative units (CADER). They are nominative and can be cashed by any banking institution. Physical and electronic devices prevent frauds. In addition to BANSEFI,[9] three commercial banks can issue PROCAMPO cheques: BANAMEX, BANCOMER and BANORTE. These are selected in each region by calls for tender. The unit cost of a cheque ranges between MXN 11.04 and 12.74 before tax, depending on the bank.

For farmers who have a bank deposit account, transferring the payment to these accounts is administratively simpler, cheaper and safer than other payment options. Initially, the account number is submitted for registration to ASERCA regional offices. ASERCA central office enters the number in its database, validates it and sends it to the banking system, which in turn validates it and informs ASERCA. For each deposit, financial dealers registered with ASERCA formalise a contract with producers and send it to ASERCA regional offices, which validate the request, calculate the amount and ask permission to transfer that amount to the deposit account. Central ASERCA grants this permission and gives instructions to the banking system to proceed with the transfer. Dealers check the deposit has been made and write an acknowledgement of receipt. Producers receive information from the dealer. They also have to register each claim at the CADER, which validates the information received on the producer. The unit cost of a deposit is MXN 4 before tax.

The use of electronic cards has been recently promoted for farmers with more than 5 hectares as an even simpler and more transparent means of transferring PROCAMPO payments; this is expected to reduce administrative costs for the CADERs. The initial cost of creating the card (MXN 10 before tax) is paid by ASERCA. No minimum deposit is required so it is accessible to producers who would not be granted a bank account. The card can be used for all the different types of payments the producer is entitled to, including payments from other programmes such as *Programa de Empleo Temporal* and marketing payments. It can also be used for national and international transfers and as a debit card without paying a withdrawal commission. In 2002, over 200 000 cards were distributed and it was expected that by 2006 all farmers will have an account. The system will be evaluated once a year by an independent office. This method was awarded a prize for innovation in reducing administrative costs.

4.6. Information technologies

ASERCA's system of geographic information (SIGA) is a technical component of PROCAMPO projects aimed at small producers. It includes the following subcomponents (ASERCA, 2002b):

- Verification that the agricultural activities being carried out by the producer receiving a payment corresponds to what has been declared to the CADERS.

- A complete register of land and producers engaged in all PROCAMPO programmes, as well as the registration of all "*ejidal*" properties. This task was finished in 2004.

- Taking satellite images to identify the technical and productive capabilities of all agricultural land.

- The identification of agricultural regions and agricultural frontiers.

- The assessment of the environmental impacts of PROCAMPO, the agriculture reform, and all agricultural activities.

- The estimation of such impacts at the national level.

In 1993, all eligible farmers were registered in the PROCAMPO database, but until 2004 they needed to register their claim for each cycle. As of 1996, no new producer could register with this programme. ASERCA checks land use declared with the information contained in the database. For environmental projects, random checks are performed.

4.7. Estimation of PRTCs for PROCAMPO

Policy-related transaction costs (PRTCs) are estimated for the two public bodies involved in the implementation of PROCAMPO payments: ASERCA and the CADERs. Estimates are based on publicly available information (i.e. made at a relatively low marginal cost). The estimation methods used for the two organisations reflect data availability. Increasing the precision of these estimates would require additional collection of information.

PROCAMPO costs must first be isolated from the PRTCs of other programmes implemented by ASERCA and the CADERs. An attempt is then made to allocate total PRTCs to the various tasks required for the implementation of PROCAMPO payments, listed in

Figure 1.1 of Chapter 1. The focus is on the tasks identified for the distribution and monitoring of payments.

ASERCA

As ASERCA is a body of SAGARPA, its total administration cost is directly available from SAGARPA's budget under "operational costs". It was projected that for 2003 this budget cost would be MXN 326 million (USD 30 million) in 2003. It covers the administration cost of central ASERCA and its regional offices. As shown in the first column of Table 4.1, the budget distinguishes:

- the cost of staff (permanent and temporary) – including wages, social security and other social transfers;
- the cost of services from banks, consultants, cleaners, etc.; and
- the cost of equipment and stationary.

Table 4.1. **Administration cost of ASERCA: Budget plan 2003**

	Total ASERCA's PRTCs	ASERCA's PRTCs on PROCAMPO (66% of total)	
	MXN mn	MXN mn	%
Labour costs	205.1	135.1	63
Service contracts	95.6	63.0	29
of which banking costs	*–*	*41.0*	*19*
Equipment and stationary	25.2	16.6	8
Total	325.8	214.6	100

Source: SAGARPA (2003a).

ASERCA's total administration cost in the 2003 budget plan is split between PROCAMPO and marketing payments, the two main programmes managed by ASERCA, on the basis of the number of professional staff (support staff is not taken into account) in each section of the organisation (see the organisation chart of central ASERCA in Figure 4.1). It is therefore assumed that the ratio of professional staff to support staff is the same in all sections of ASERCA. This allocation is also assumed to be the same in central ASERCA and the nine regional offices, which could not be isolated in the budgetary statistics.

Allocation is done in two steps. First, ASERCA's total operational cost is allocated to the various sections according to their share of professional staff. Second, the different sections need to work with one or the other programme, both or none. Some sections of ASERCA (on the left side of Figure 4.1) are solely in charge of the implementation of marketing payments (the General Co-ordination of Marketing and the Directorate for Marketing Payment Delivery) and their cost share can therefore be allocated to this programme. In the same way, other sections of ASERCA (in the middle of Figure 4.1) such as most Directorates General under the General co-ordination of support[10] (with the exception of the Directorate for Marketing Payment Delivery) can be directly related to PROCAMPO. However, the tasks of the Head Directorate (excluding the unit for agri-food studies and support to international trade negotiations, which is not allocated to either of the programmes) and the DG for administration and finance are shared between PROCAMPO and marketing payments. The costs of the shared sections are then split between the two programmes based on the number of farmers receiving payments (which is a proxy for the number of applications). The assumption is that applications for PROCAMPO or marketing payments have equal processing time. As more farmers receive PROCAMPO payments than do marketing payments, 98% of the shared costs are allocated to PROCAMPO. In the end, PROCAMPO is estimated to account for 66% of the administration cost of ASERCA (Figure 4.2), i.e. MXN 215 million (USD 20 million) (second column of Table 4.1).

Figure 4.2. **Allocation of ASERCA's PRTCs to PROCAMPO**

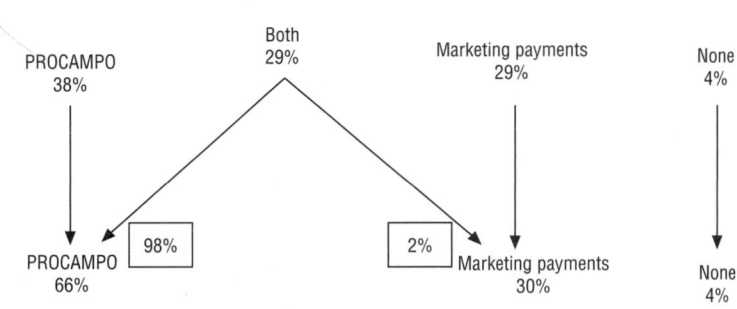

Source: Secretariat's estimate based on ASERCA's flow chart.

This PRTC of MXN 215 million is broadly allocated to the different tasks performed by ASERCA when implementing PROCAMPO, i.e. the identification of beneficiaries, the processing of applications, actual payment, checking of eligibility and compliance, and monitoring and co-ordination of the delivery system and the whole administration system. As for the attribution of costs to PROCAMPO, the allocation by task is carried out on the basis of the number of professional staff in each section of ASERCA, specific tasks having been identified for specific sections as described below. The estimated cost allocation is shown in the first column of Table 4.2.

The main task of ASERCA is to **monitor** and control the administration of the system. This role is allocated to the Head Directorate and the DG for administration and finance, but also to the DG for direct payment programming and evaluation and the Directorate for budgetary control and monitoring of direct payments of the payment DG. Under the general co-ordination of support, the DG of means of payments is mainly involved with **delivering actualpayments**, while the **processing** of payments is done in the DG of information systems for direct payment delivery. The DG of geospatial analysis is given the task of checking **eligibility and compliance**. Banking costs for actual payment delivery are part of the ASERCA budget, but cannot be identified precisely as contracting costs of ASERCA in the budget include more than just payments to banks. However, total banking costs for PROCAMPO cheques were estimated by the banks to be MXN 41 million (USD 3.8 million) in 2002 (Table 4.1). This amount is directly allocated to the task of proceeding to "actual payment" and deducted from the total amount of ASERCA administration costs to PROCAMPO, to be allocated to all tasks.

ASERCA also plays a role in the design and evaluation of PROCAMPO. The Directorate for planning and development of the DG of information systems for direct payment delivery is involved with the **design** of the programme. **Evaluation** of the programme is carried out by the directorate for studies and analysis of rural support of the DG of computer systems, while other parts of this DG are dedicated to monitoring the delivery

Table 4.2. **Allocation of PROCAMPO's PRTCs**

Tasks	PRTCs of ASERCA	PRTCs of CADERs	Total PRTCs	PRTCs as a % of payments	PRTCs per producer	PRTCs per hectare
	MXN mn	MXN mn	MXN mn	%	MXN	MXN
Design[1]	5	0	5	0.04	2	0
Evaluation	6	0	6	0.05	2	0
Identification of beneficiaries	0	48	48	0.37	17	4
Processing of applications	30	68	97	0.75	34	7
Actual payment	46	25	71	0.54	25	5
Eligibility/compliance	13	23	36	0.28	13	3
Monitoring[2]	115	0	115	0.88	40	8
Total	215	164	379	2.90	133	27

1. Does not include SAGARPA costs other than ASERCA's.
2. Includes management, computer operation, and organisation costs.
Source: Secretariat's estimates.

system. The collection of information at the ASERCA level is not considered as a specific PRTC given that it is generated from the implementation process which requires that information from applications be centralised in a database managed by ASERCA. This could be interpreted as a benefit of PROCAMPO implementation.

CADERs

The CADERs administration costs are not identified in the SAGARPA budget. The estimate of the costs they incur when implementing PROCAMPO is based on labour costs only. For each region, the number of staff in CADERs, by administrative category, and the monthly wages of civil servants of the different categories are available from SAGARPA. Labour costs of CADERs by region are obtained by multiplying the number of staff in one category by the average wage of that category and by summing the labour costs across all categories of staff. Total labour costs of CADERs amounted to MXN 644 million (USD 60 million) in 2002.

CADERs implement PROCAMPO as well as many other programmes. It is therefore necessary to estimate the share of their labour costs spent on PROCAMPO implementation in order to obtain PROCAMPO PRTCs from CADERs. This is done by using the number of days CADER employees spend on PROCAMPO tasks. This information is available from a SAGARPA report (2003b) on PROCAMPO management indicators for the spring-summer agricultural cycle of 2002. The duration of each task necessary to implement PROCAMPO payments is given by the CADER and the region. It is used as an approximation of the time spent on PROCAMPO tasks. For the autumn-winter cycle, information on total days spent on PROCAMPO is available and allocated arbitrarily to individual tasks using information from the spring-summer cycle. Table 4.3 shows the number of days allocated to PROCAMPO tasks on average, but the calculation was done for all regions as labour time and cost vary by region with the number of applications and the staff composition of CADERs. In total, CADERs' employees spend over 90 days on PROCAMPO: 43 for the spring-summer cycle and 48 for the autumn-winter cycle (average of 2001/02 and 2002/03). This represents about a quarter of their time.[11] As a result, average PRTCs of PROCAMPO from CADERs is estimated to be MXN 164 million (USD 15 million) (Table 4.2).

For each region, labour costs are finally allocated to specific tasks on the basis of the number of days spent on both PROCAMPO cycles in each region. The result of this

Table 4.3. **Average number of days CADERs spend on PROCAMPO**

	Spring-Summer 2002	Average Autumn-Winter 2001/02 and 2002/03	Total 2002[1]	Total 2002 in percentage terms
Identification of beneficiaries	14	n.a.	29	8
Processing of applications	18	n.a.	37	10
Register the application	6	n.a.	14	4
Control and calculation	6	n.a.	12	3
Printing	5	n.a.	11	3
Actual payment	5	n.a.	10	3
Eligibility/compliance	7	n.a.	14	4
Total labour days on PROCAMPO	43	48	91	25

n.a.: not available.
1. The Autumn-Winter cycle is allocated to individual tasks on the basis of the Spring-Summer allocation.
Source: SAGARPA (2003b) and ASERCA (2003).

calculation provides an estimate of CADERs' PRTCs on PROCAMPO for specific tasks (Column 2 of Table 4.2). It is then added to the PRTCs of ASERCA, by task, to obtain the cost of implementing and controlling PROCAMPO payments (Column 3 of Table 4.2).

Total PROCAMPO PRTCs are estimated at MXN 379 million (USD 35 million), or less than 3% of the value of total payments. To give a more precise picture of their relative size, PRTCs are also related to the number of producers receiving payments and hectares covered. Average estimated PRTCs per producer are MXN 133 (USD 12.4) for an average payment of MXN 4 592 (USD 427). The average payment per ha is MXN 947 (USD 88) for an estimated PRTC of MXN 27 (USD 2.5). Given the high number of farmers receiving payments (Table 4.4), the PRTCs of PROCAMPO are relatively modest, because transactions are relatively standard and most tasks are computerised.

Table 4.4. **PROCAMPO transfers in 2003**

	Number of producers	Payments	Area	Number of farms
	'000	MXN mn	'000 ha	'000
Autumn-Winter cycle 2003 fiscal year	438	2 711	3 105	587
Spring/Summer cycle 2003	2 405	10 343	10 681	3 520
Total	2 843	13 054	13 786	4 107

Source: ASERCA (2004), Table 9.

4.8. Conclusions

The highest PRTC comes from the monitoring of the whole system, *i.e.* the administration and coordination of the whole system by the Head Directorate and DG for the administration and finance of the central ASERCA office. Although ASERCA deals mostly with PROCAMPO, it increasingly carries out other tasks which were not taken into account in this case study and could account for part of such shared costs. The second highest cost comes from the processing of applications, two-third of which occurs at the local level. Identification of the beneficiary has a relatively low cost as all PROCAMPO producers are registered in a database. Similarly, this database is used to check eligibility, at a relatively low cost, through the monitoring of successive applications and GIS information. In addition, it is expected that the cost of proceeding with payments will decrease as more producers cash their payment by using a debit card.

PROCAMPO PRTCs were estimated using publicly available information. A number of assumptions had to be made regarding the allocation of costs by programme and task, and the value of labour costs. More refined estimates could be found with more detailed information, but the current method was implemented at a relatively low cost.

PROCAMPO administration costs are relatively modest given the large number of farmers that receive payments, which is larger than the number of farmers that benefited from the former price support system. The low integration of some farmers in the banking system has also been a challenge. As withdrawal cards develop and participants are no longer required to register for each payment cycle, costs are likely to decrease. It would be interesting to follow the evolution of PROCAMPO PRTCs as the different ways of delivering payments develop and the facilities put in place to administer PROCAMPO (withdrawal cards, database and GIS system) are used for other programmes. It is expected that PRTCs will further decrease as the delivery and monitoring network, and the information systems, are shared.

Notes

1. Programme of Direct Assistance to the Countryside (*Programa de Apoyos Directos al Campo*).

2. National Basic Food Company (*Compania Nacional de Subsistencias Populares*).

3. Most other payments under the ALIANZA programme consist of subsidies on inputs (mainly investments) or on-farm services.

4. There are 32 administrative regions (States) in Mexico.

5. From 2005, programme participants are no longer requested to register claims for each cycle.

6. The implementation system of capitalised PROCAMPO is different from the payments types because of the nature of the assistance. CADERs check the request, receive and register the economic project; ASERCA produces and prints the official document and registers the economic project and the credit. A State committee evaluates the project's technical feasibility. A financial institution evaluates the financial viability of the project and formalises the credit.

7. The Inter-American Development Bank, which funds part of capitalised PROCAMPO, will evaluate the programme.

8. Bearer cheques can be directly cashed by the bearer.

9. BANSEFI (*Banco del Ahorro Nacional y Servicios Financieros*) is a development bank whose purpose is to promote saving culture and to provide technical assistance and financial services to the popular credit and saving entities of the Federal Government.

10. The payment DG (with the exception of the Directorate for marketing payment delivery), the DG for direct payment programming and evaluation, the D.G. of information systems for direct payment delivery, the DG of computer systems and the DG of geospatial analysis.

11. CADERs estimate their work on PROCAMPO accounts for 80% of their time. If this high estimate was used, PROCAMPO's total PRTCs would be double at MXN 730 million or 5.6% of transfers.

References

ASERCA (2002a), "PROCAMPO", presentation by the Directorate General of Means of Payments, General Co-ordination of Support, December.

ASERCA (2002b), "*Reporte BID – Evaluacion ambiental del PROCAMPO*", presentation by the General Co-ordination of Support, Directorate General of Means of Payments, December.

ASERCA (2003), *Informe PEF 2003*, Directorate General of Information Systems for Direct Payments Delivery, General Co-ordination of Support.

ASERCA (2004), "*Comité de Control y Auditoría en ASERCA*", Control and audit Committee, first ordinary session 2004, Directorate General of Means of Payments and Directorate General of Information Systems for Direct Payments Delivery, March.

OECD (2003), *Agricultural policies in OECD countries: Monitoring and Evaluation*, OCDE, Paris.

SAGARPA (2003a), *Resumen del estado del ejercicio prepuestal 2003*, Oficialia mayor, Dirección general de eficiencia financeria y rendición de cuentas, (Summary of the state of the budgetary exercise 2003, Main office, Directorate General of financial efficiency and accounting), avril.

SAGARPA (2003b), *PROCAMPO indicadores de gestión, cyclo agricole Primavera-verano 2002, Duración del processo operativo por Delegación y CADER* (Management indicators, Spring/Summer cycle – Time spent on managing payments by delegation and CADER), March.

ISBN 978-92-64-03091-6
The Implementation Costs of Agricultural Policies
© OECD 2007

PART II

Chapter 5

A Case Study of the Policy-related Transaction Costs of Direct Payments in Switzerland

Executive Summary

This case study presents an estimate of the transaction costs generated by the Swiss direct payment system. The costs are estimated for two case studies concerning the cantons of Grisons and Zurich, for which the implementation and control costs are assessed at five levels; namely the State, the canton, the control organisations, the borough, and the farm. While the costs accruing to public authorities and control organisations can be determined with exactitude, the numerous factors which influence costs as well as the differences between farms result in uncertainties when assessing labour expenditure and labour costs at farm level.

Discussions about the efficiency of direct payments and agricultural policy measures form the background of this case study. To date, transaction costs have been neglected in investigations on efficiency. However, due to the new orientation of agricultural policy in Switzerland and the European Union, with the associated de-linking of measures from production levels and their link-up with specific or farm-related regulations (cross compliance), the extent to which transaction costs influence the implementation and efficiency of policy and farm participation must be determined.

In Switzerland, direct payments have gained considerably in importance, with regular development in the delivery system and regular increases in the funding. between 1992 and 1998. Based on a new article to the Constitution and new agricultural laws, since 1999 direct payments have been linked to proof of ecological performance. General direct payments (Red Ticket measures) remunerate the multi-functional services provided by agriculture, while special ecological and ethological services (Green Ticket measures) are covered by additional payments.

In 2003, agriculture received a total of CHF 2.47 billion in the form of direct payments; 81% of this sum was granted for general payments and 15% for ecological direct payments. The remaining 4% covered summer pasturage contributions. In the same year, 89% of all Swiss farms fulfilled the conditions of proof of ecological performance. These farms covered 96% of the total utilised agricultural area. At the same time, organic farms held a 10% share, and this share rose noticeably in the 1990s. Participation also increased strongly with regard to ethological payments: in 2003, 30% of all livestock units were kept in accordance with BTS (animal housing system) requirements and 62% of total livestock units were entered in the RAUS (outdoors regularly) programme.

Overall transaction costs amount to CHF 3 million in Canton Grisons and CHF 3.9 million in Canton Zurich. Given the total transaction costs of CHF 1 100 per farm, costs between CHF 690 and CHF 755 arise for the farms. In Canton Grisons, these values are set off by overall direct payments amounting to CHF 167 million or CHF 60 800 per farm, and in Canton Zurich to CHF 141 million or CHF 38 500 per farm. This results in high transfer efficiency for the payments. In Canton Grisons, total transaction costs amount to 1.8% of the overall direct payments and 2.8% in Canton Zurich.

Public authorities pay approximately 37% of the total transaction costs in Canton Grisons and approximately 30% in Canton Zurich. The farms cover the remaining costs. From the point of view of the public authorities, the relationship between transaction costs and the direct payments disbursed can be regarded as very efficient. Transfer efficiency is influenced by factors relating to the system and environmental conditions as well as by the overall direct payments disbursed.

Overall transaction costs depend on five factors: 1) farm size; 2) the farm's participation in Green Ticket measures (ecological and ethological programmes); 3) organisation differences between the two cantons; 4) orientation of the farms; and 5) environmental influences. On the other hand, transaction costs per participating unit depend primarily on the size of the farm. The larger the farm, the lower the transaction costs per area unit, as bigger farms can spread their fixed cost share of the transaction costs over a greater area. The farm's fixed cost share and the transaction costs per farm depend not only on the processes stipulated by the State, but also on the capabilities of the farm manager.

The classification and interpretation of transaction costs must be in direct relation with the respective direct payment programmes and agricultural policy target system. Under the Swiss system, the services agriculture is called upon to provide as defined in the Federal Constitution are remunerated by means of direct payments. This means that transaction costs can be interpreted as part of the costs of quality assurance. The greatest part of the costs is attributable to controlling the regulations governing eligibility to receive payments. Thus, in the first instance, the sum of the direct payments is attributable to the desired quality of public goods, i.e. the multifunctional services provided by agriculture. This applies to both public authorities and the farms. Within the scope of the current direct payment system, any substantial reduction of transaction costs can probably only be achieved by modifying the quality requirements of the multifunctional services. An improvement in implementation and control efficiency demands simultaneous optimisation of transaction costs and the quality of the services, whereas these two dimensions exhibit conflicting objectives.

Abbreviations

German		English	
ÖLN	Ökologischer Leistungsnachweis	PEP	Proof of ecological performance
		PRTC	Policy-related transaction costs
RGVE	Raufutter erzehrende Grossvieheinheit	GLU	Grazing livestock unit
TEP	Tierhaltung unter erschwerten Produktionsbedingungen	TEP	Keeping livestock under difficult conditions
BTS	Besonders tierfreundliche Stallhaltungssysteme	BTS	Animal housing systems
RAUS	Regelmässiger Auslauf im Freien	RAUS	Turning animals outdoors regularly
Bio	Biologischer Landbau	Bio	Organic crop farming
		PSE	Producers support estimate
IP	Integrierte Produktion	IP	Integrated crop production
GVE	Grossvieheinheit	LSU	Livestock unit
A	Are	A	Are
Ha	Hectare	Ha	Hectare
BLW	Bundesamt für Landwirtschaft	FOAG	Federal Office for Agriculture
ALN	Amt für Landschaft und Natur Zürich		Zurich Office for Landscape and Nature

5.1. Background and goal

This case study examines the implementation of the direct payment system in two Swiss Cantons and estimates associated costs. The primary objective is to evaluate transaction costs, giving special consideration to cross compliance under the Swiss direct payment system. Due to the relationship between the regulations governing the proof of ecological performance for the farm's eligibility to receive direct payments and the ecological compensation programmes, the latter is also taken into account in the evaluation. Since the sums paid out in association with the ecological quality regulations are virtually insignificant, these measures are not included.

The primary objective can be sub-divided into three sections, or tasks:

- representation of the current direct payment system and the individual payments, giving due consideration to the regulations relevant for the farms, overall cross compliance on the farms, as well as the existing implementation and control institutions (Section 5.2);

- selection of suitable methods for the evaluation of transaction costs (Section 5.3); and

- evaluation of transaction costs under the existing direct payment system at federal and cantonal levels, as well as at the level of the implementation and control institutions and the farms (Section 5.4).

The evaluation of transaction costs presented in this chapter is limited to those costs arising from the implementation and control of the respective policy measures. In particular, this means that the costs arising from the following policy programmes are not taken into account.

- costs for planning and setting up a policy programme; and

- farm modification costs for participation in a programme (costs associated with a switch-over, reduction of number of livestock units or changes in use of land to comply with PEP, etc.).

5.2. The Swiss direct payment system

This section explains the direct payment system applied under Swiss agricultural policy. A short section describes the creation of this system. It is followed by a definition of its position within the constitutional context and its role in agricultural policy objectives. As direct payments are subject to numerous regulations it is necessary to describe them in detail to provide a basis for the calculation and interpretation of the transaction costs. The development of the direct payments granted and the areas, or animals, covered by the programmes is then presented.

Background of the direct payment system

In 1992, the seventh Federal Agricultural Report heralded a change in Swiss agricultural policy. Up until that time, policy had focussed mainly on ensuring that the nation was supplied with essential goods and services, and agricultural support was realised largely by means of market intervention. At the beginning of the 1990s, three reasons arose which, in parallel to the on-going GATT negotiations, demanded a fundamental revision of Swiss agricultural policy (Rieder, 1998):

- over-production resulting from the high price policy was costly for the State;

- high degree of pollution; and

- deteriorating income distribution between large and small farms, and likewise between lowland and mountain farms.

To counter this, a new agriculture article was introduced in 1996 into the Federal Constitution defining a policy based on the idea of multifunctional agriculture. This new definition is based on the perception that agriculture not only produces foodstuffs but also provides public goods which cannot be remunerated via the market. The provision by agriculture of these public utility services is ensured and remunerated by means of direct payments which are not linked to production.

Following the introduction of the first direct payments in the 1970's, the system underwent systematic development between 1993 and 1999, whereby the following four steps were fundamental:

1. The seventh Agricultural Report of 27 January 1992 formed the basis for the separation of price and income policies, and for the introduction of direct payments which were not linked to production as per 1 January 1993. New ecological direct payments were introduced. These were subject to regulations which related either to the area or animals involved (cross compliance on the level of farm activities).

2. Passage of the new constitutional article 104 BV (Federal Constitution) following the referendum of 9 June 1996. This article forms the basis for the agricultural policy 2002 and thus for the new law on agriculture.

3. Complementary direct payments linked to a minimum share of ecological compensatory areas of 5% as per 1997 and 7% as per 1998.

4. The new law on agriculture and the present direct payment system came into force on 1 January 1999. Entitlement to direct payments is linked to the proof of ecological performance for all farms (cross compliance on the farm level).

The Swiss direct payment system as an element of agricultural policy

The new agricultural policy, and thus the direct payment system, is based on Article 104 of the Swiss Federal Constitution. Section 5.1 of this chapter defines the multifunctional role of agriculture:

- Guaranteed food supply for the population: ample food supplies should not only be available for the population in "normal times", but it must also be possible to step up production to an adequate level in times of crisis.

- Conservation of natural resources and upkeep of rural scenery: the "soil" factor is of primary importance for agricultural production. As such, the protection of soil fertility is the central element in the conservation of natural resources. Furthermore, water, air, fauna and flora are also natural resources and must be given due consideration in the course of agricultural production.

- Rural scenery is marked by agricultural production. Both settlement structures and farming practices produce typical landscapes. However, husbandry does not involve the maintenance of a particular state, but rather the avoidance of disruptive interventions and influences.

- Decentralised settlement of the country: village communities, with their specific political and cultural life, should be preserved and developed thanks to strong agriculture.

Under the terms of the Federal Constitution, the State must ensure that these assignments are fulfilled by means of sustainable and market-orientated production. Due consideration must be given to the fact that, to a large extent, the tasks assigned to agriculture under the Constitution tie-in directly with prevailing production. There is a trade-off relationship between production targets and the environmental targets set for agriculture: the more intensive the production of agricultural goods, the greater the pollution. Therefore, Swiss agricultural policy perceives the target of sustainability mainly in its ecological dimension, but at the same time, production must be competitive and should allow farmers to produce efficiently and meet current demands.

Section 3a of the Constitution permits the State to supplement farm incomes by means of direct payments. The payments are made as remuneration for the provision of multifunctional services on condition that ecological performance is proven. The upper part of Figure 5.1 shows the direct payment measures as foreseen by law and their significance for the fulfilment of the multifunctional tasks assigned to agriculture.

There is a difference between general direct payments and ecological direct payments which are based on Section 3b of Article 104 of the Constitution. Under the terms of this article, the State is required to grant economically attractive incentives to encourage forms of production that are particularly environmentally acceptable, animal-friendly and close to nature. This means that specific direct payments can be granted for the provision of additional services as illustrated in the lower part of Figure 5.1. These ecological direct payments encourage additional, clearly defined services. While it is not possible to identify with certainty the effects of general direct payments on the fulfilment of the multifunctional tasks assigned to agriculture, the benefits of ecological direct payments which remunerate a specific service are immediately obvious.

Regulations of the direct payment system

In Switzerland, the disbursement of direct payments to farms is subject to regulations, whereby there are clear distinctions between the regulations for general direct payments and ecological or ethological contributions.

Cross Compliance Concept

In the relevant literature, the tie-in between financial support – in the case of agriculture, the eligibility to receive direct payments – and specific (ecological or social) regulations is generally referred to as "cross compliance": "*In the European Union debate, the terms cross compliance and environmental conditionality are often used interchangeably to describe the linking of a farmer's eligibility for agricultural subsidies to environmental conditions*" (European Environment Agency 2004). In this way, agricultural and environmental targets are linked together.

In the cross compliance concept, the differentiation between general and ecological direct payments can be made using the Red and Green Ticket Approaches (Christensen, 2000):

- Red Ticket Approach: if a farmer does not provide the required services, the payments are lowered or completely discontinued. The payments are linked directly to agricultural policy targets and only depend to a small extent on the benefits or costs of a programme.
- Green Ticket Approach: farms are offered an additional payment for measures which exceed the minimum requirements. To be precise, this is no longer a case of cross compliance, but rather an ecological requirement.

Figure 5.1. **The direct payment system**

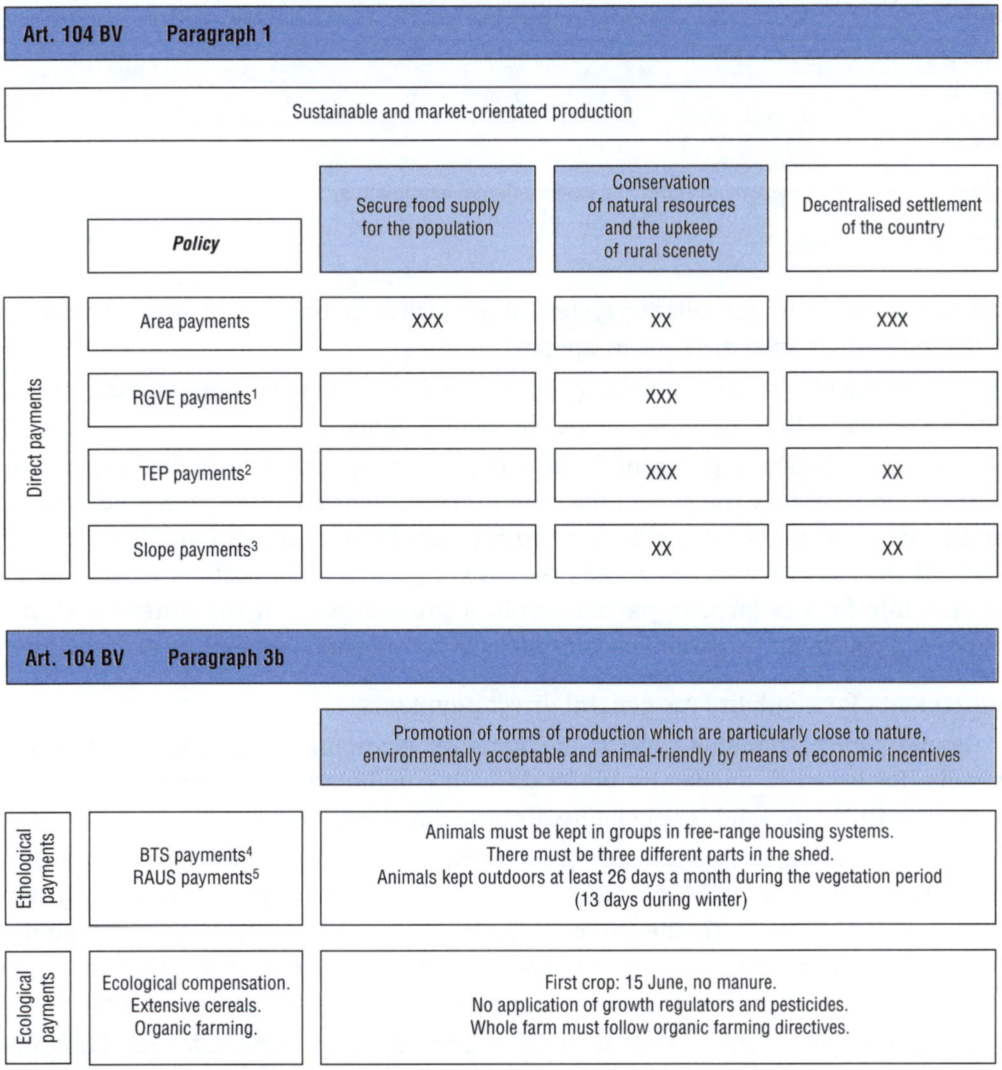

Art. 104 BV	Paragraph 1			
		Sustainable and market-orientated production		

Policy	Secure food supply for the population	Conservation of natural resources and the upkeep of rural scenety	Decentralised settlement of the country
Area payments	XXX	XX	XXX
RGVE payments[1]		XXX	
TEP payments[2]		XXX	XX
Slope payments[3]		XX	XX

(Direct payments)

Art. 104 BV	Paragraph 3b

Promotion of forms of production which are particularly close to nature, environmentally acceptable and animal-friendly by means of economic incentives

Ethological payments	BTS payments[4] RAUS payments[5]	Animals must be kept in groups in free-range housing systems. There must be three different parts in the shed. Animals kept outdoors at least 26 days a month during the vegetation period (13 days during winter)
Ecological payments	Ecological compensation. Extensive cereals. Organic farming.	First crop: 15 June, no manure. No application of growth regulators and pesticides. Whole farm must follow organic farming directives.

1. Payments for keeping grazing farm animals.
2. Payments for keeping livestock under difficult conditions.
3. Payments for farming on steep slopes.
4. Payments for animal housing systems.
5. Payments for turning animals outdoors regularly.

Source: FOAG.

The difference between general and ecological direct payments is to be found in the terms of payment. If a farmer fails to fulfil the requirements of proof of ecological performance, or only fulfils them in part, his general direct payments are lowered. On the other hand, he receives no ecological direct payments at all if he fails to fulfil all the requirements of the associated regulations.

Table 5.1 illustrates the two categories of cross compliance: Farms must provide an ecological performance to the extent of x. The costs of observing this requirement amount to C, direct payments amounting to y are granted for the performance of the service. Additional (environmental) services are rewarded with a payment z. If the requirements are not fulfilled, payments are cut by the factor α.

Table 5.1. **Definition of cross compliance measures**

Policy type	Net income with compliance (doing x or more)	Net income without compliance (not doing x)
Red Ticket	$y - C$	$(1 - \alpha)y$
Green Ticket	$z - C$	0

y: Payments to agriculture.
C: Costs of observing environmental requirements.
z: Support in case of fulfilment of an additional environmental requirement.
α: Reduction of payment in case of non-observance of the requirements: $(1 > \alpha > 0)$.
Source: Christensen (2000, p. 8).

Based on the cross compliance system, it is possible to draw conclusions regarding the anticipated participation of farms in the programmes:

- Payments linked to production: the less dependent a payment is on production, the more likely a farmer is to react to the ecological requirements.

- Amount of the direct payment in relation to the costs of participation: a farm's participation costs are the sum of the minimum proceeds resulting from production, the additional expenditure arising from the fulfilment of the regulations and the transaction costs. If the direct payment exceeds the farm's participation costs, it is economically worthwhile for the farm to participate in a programme. On the other hand, if the opposite applies, participation leads to a loss of income and is therefore not worthwhile.

Requirements for eligibility for general direct payments

General direct payments cover area payments, payments for grazing livestock units, payments for livestock husbandry under difficult conditions and slope payments. Three types of condition are valid for eligibility of these direct payments:

- General type of requirements: only those farm managers who run a private farm and have their place of residence in Switzerland are entitled to receive direct payments. Farms owned by the State, the cantons, the boroughs or legal entities receive no direct payments. In addition, farms which overstep the regulations stipulating the highest permissible number of livestock units do not receive any direct payments either.

- Proof of ecological performance is another principal requirement for receiving direct payments and forms a link between agricultural and environmental policy targets.

Proof of ecological performance consists of five elements (BLW, 2004):

- animal-friendly husbandry of livestock and observance of animal protection laws;
- balanced nutrient/fertiliser balance sheet;
- adequate share of ecological compensatory areas;
- regulated crop rotation;
- suitable soil conservation measures from arable zone up to and including mountain zone I;
- limited choice and regulated use of crop treatments and consideration of pollution thresholds and forecasts.

Structural and social requirements: structural requirements cover the criteria size of farm, minimum labour requirement, on-farm workforce and age of the farm manager. In addition, general direct payments are limited according to the size of the farm, the number of animals as well as income and assets. Figure 5.2 illustrates the grading according to size of farm and numbers of livestock.

Figure 5.2. **Grading of contributions according to area and number of livestock**

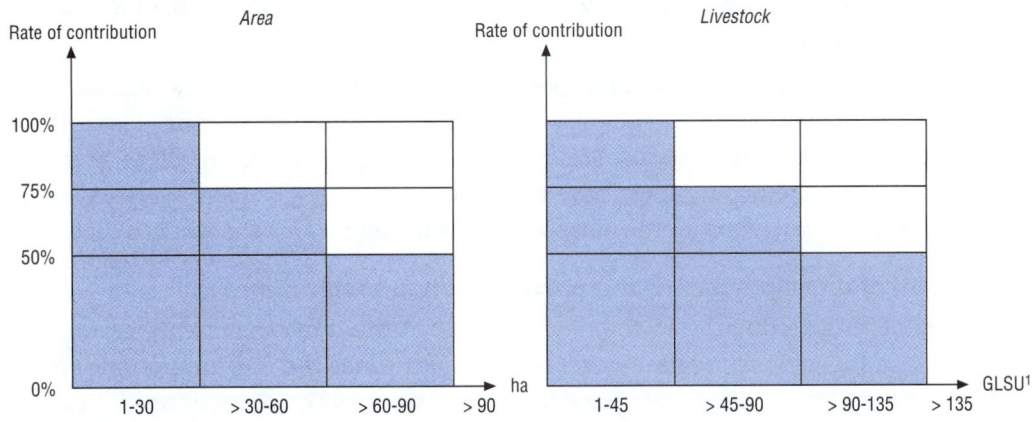

1. Grazing livestock unit (GLSU).

Source: Agricultural Report (BLW, 2003b).

In addition to proof of ecological performance and observance of general structural and social stipulations, eligibility to receive direct payments also depends on adherence to regulations specific to agriculture as defined in the laws on water protection, pollution control, nature conservation and protection of rural landscape. If a farmer contravenes these laws, he is not only fined but direct payments made to him can also be withheld. Furthermore, as of 2007 eligibility for the receipt of direct payments also depends on proof of basic professional training in agriculture.

Requirements for receipt of ecological contributions

Ecological contributions cover payments for ecological compensation, the extensive production of cereals and rapeseed as well as for organic farming. As a whole, the requirements for eligibility for these direct payments can be regarded as utilisation restrictions. As discussed above, these contributions are designed to compensate for the yield losses and extra outlay arising from participation.

Fertiliser and utilisation restrictions (due date for mowing) are, for example, relevant in the case of extensive meadowland and fallow, while the prohibition of growth regulators, fungicides or insecticides plays a role in the extensive production of cereals and rapeseed. The cultivation of high-standing trees or the maintenance of hedges or dry-stone walls is also rewarded by ecological direct payments.

Eligibility for receipt of contributions for organic farming is subject to the fulfilment of the stipulations of the Federal bio-regulations, whereby the salient point is that the entire farm must be run according to the guide-lines for organic farming. In particular, these include the prohibition of the use of mineral fertilisers, the foregoing of the use of all forms of chemical additives, and the observance of more stringent animal husbandry regulations.

Requirements for receipt of ethological contributions

Ethological contributions cover payments for particularly animal-friendly housing systems (BTS) and turning the animals outdoors on a regular basis (RAUS):

- An animal housing system is regarded as particularly animal-friendly when it consists of several areas where the animals are kept free in groups, when there is an adequate source of daylight and the animals have suitable opportunities to rest, move about and occupy themselves according to their natural behaviour patterns. Stipulations for the various types of livestock are set down in a special regulation for particularly animal-friendly housing systems.

● Regular turning out means that grazing livestock are turned out on meadowland for at least 26 days per month during the vegetation period and that they are outdoors on at least 13 days per month during the winter season. Pigs must be able to go outdoors on at least three days per week. Rabbits and poultry must have the opportunity to go outdoors every day. In this case too, further stipulations for the individual types of animal are set down in a regulation.

Development of direct payments and participation in programmes

By way of an introduction to the development of the direct payments granted and participation in the various programmes, Figure 5.3 illustrates the most important types of direct payment and the changes they have undergone since 1993. Contributions for summer pasturing and payments for the cultivation of arable land are not included in this diagram:

Figure 5.3. **Development of the Swiss direct payment system**

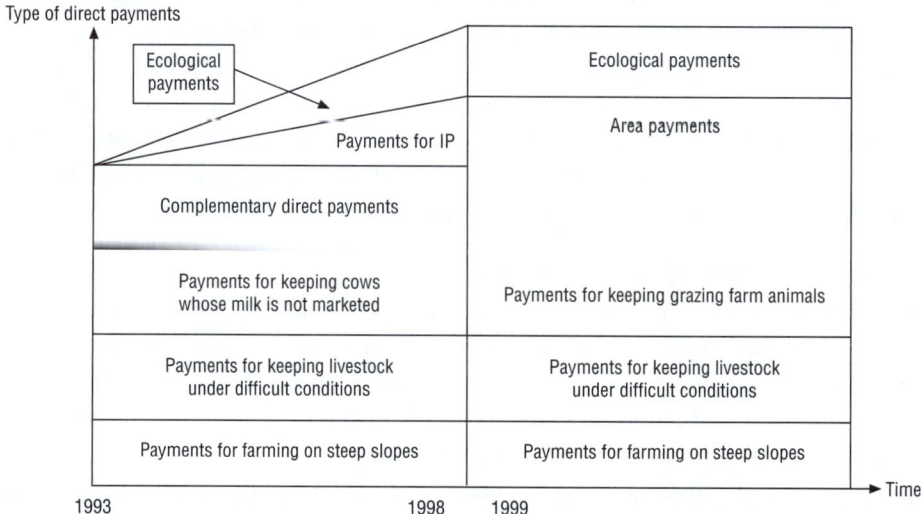

IP = Integrated production.
Source: Direct payments to agriculture 1998 (BLW, 1998, p. 61).

There has been a marked decline in the importance of market support in Switzerland due to the agricultural reform and the development of the direct payment system under which price and income policies are no longer interdependent. Prior to 1992, the share of market support in the total PSE (Product Support Estimate) was 78% and still amounted to 58% in 2002 (OECD, 2004). On the other hand, direct payments have become much more important (Rieder *et al.*, 2003). This is illustrated in Figure 5.4, which shows the development of the direct payments granted from 1993 onwards. In addition to the rise in total payments, the change of system in the year 1999 can be seen quite clearly.

A main feature of the new agricultural law was the coupling of direct payments to the fulfilment of proof of ecological performance. At the same time, the complementary direct payments, IP contributions and a part of the payments for organic farming were converted into area payments. Thus, the decline in the contributions for organic farming is a result of the system; however, the development of organic farming was not affected by the change of system.

Figure 5.4. **Development of direct payments since 1993**

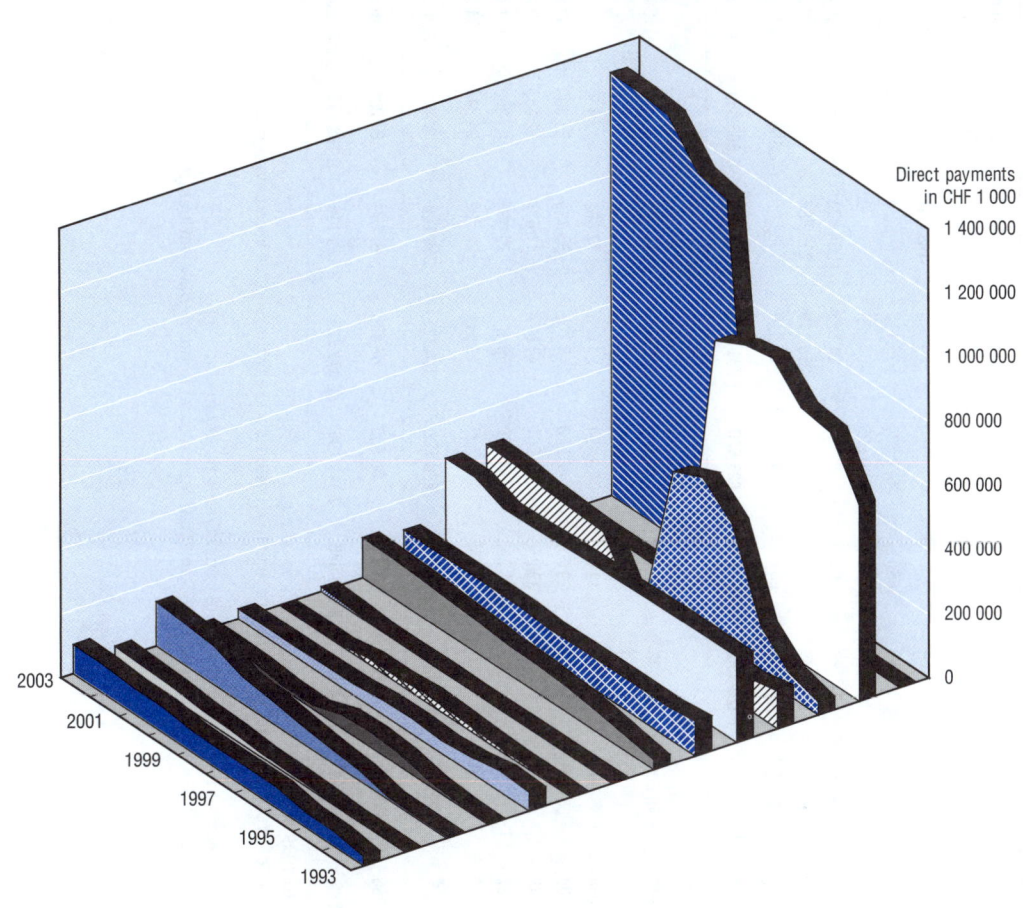

Source: Table 5.2.

By the year 2001, there was a slight decline in contributions for the keeping of livestock under difficult production conditions. This can be explained by the fact that the number of farms entitled to these payments fell due to farmers leaving this sector. However, as the remaining farms usually keep more animals than the maximum entitlement limit of 15 LSUs, as a rule it was not possible to disburse more payments in spite of the increasing area involved. After 2002, when the limit for contributions for the keeping of livestock under difficult conditions was raised from 15 to 20 LSUs per farm, a total of roughly 80 000 additional LSUs (+17%) became eligible for payments, which explains the marked expansion of these contributions.

Table 5.2. **Development of direct payments between 1993 and 2002 (in CHF 1 000)**

Direct payments[1]	1993	1994	1995	1996	1997	1998	1999	2000	2001	2002	2003
General direct payments											
Area payments (since 1999)	0	0	0	0	0	0	1 163 094	1 186 770	1 303 881	1 316 183	1 317 956
Complementary direct payments (until 1998)	610 724	779 802	794 815	888 747	872 324	825 113	0	0	0	0	0
Payments for integrated crop production (until 1998)[2]	41 550	69 652	156 412	417 253	500 925	460 020	0	0	0	0	0
Payments for keeping grazing farm animals	119 207	99 330	101 790	96 970	93 383	91 863	254 624	258 505	268 272	283 221	287 692
Payments for keeping livestock under difficult conditions	266 535	266 894	268 278	265 945	261 918	259 119	255 882	251 593	250 255	289 572	287 289
Payments for farming on steep slopes	109 147	99 297	98 860	98 670	98 070	95 110	105 207	106 790	106 686	105 862	106 154
Total of general direct payments	1 147 163	1 314 975	1 420 155	1 767 535	1 826 620	1 731 225	1 778 807	1 803 658	1 929 094	1 994 838	1 999 091
Ecologically motivated payments											
Payments for ecological compensation	31 919	48 696	66 596	79 271	87 665	90 673	100 674	108 130	118 417	122 347	124 927
Payments for ecological quality	0	0	0	0	0	0	0	0	0	8 934	14 638
Payments for green fallow[3]	3 054	5 695	8 109	12 675	19 494	24 613	17 652	17 150	0	0	0
Payments for extensive cereal and rapeseed cultivation	58 168	65 486	48 500	39 680	47 570	45 700	35 135	33 398	32 526	31 938	31 255
Payments for organic crop farming[4]	3 945	5 702	14 096	39 206	47 501	44 077	11 637	12 185	23 488	25 484	27 135
Payments for turning animals outdoors regulary	5 387	7 007	8 833	31 708	44 370	56 421	72 688	83 370	121 421	131 655	140 106
Payments for animal housing systems	0	0	0	6 025	9 523	12 641	21 002	24 748	34 034	39 029	43 257
Total of ecologically motivated payments	102 473	132 586	146 134	208 625	256 123	274 125	258 788	278 981	329 886	359 387	381 318
Payments for summer pasturing	30 750	46 630	47 830	66 910	66 553	66 885	67 571	81 238	80 524	89 561	91 381
Total	1 280 386	1 494 191	1 614 119	2 043 077	2 149 296	2 072 235	2 105 166	2 163 877	2 339 504	2 443 786	2 471 790

1. According to direct payment regulation; measures prior to 1999 are subject to subsequent measures.
2. Belonged to ecological contributions, after the introduction of proof of ecological performance IP payments were disbursed via area payments.
3. Expiring regulation of 1999-2000.
4. From 1999, the contribution for organic farming is lower as a part of it was converted into general area payments.

Source: 1993 to 1998: Report on the disbursement of direct payments (BLW, various years); 1999 to 2003: Agricultural Reports (BLW, various years).

Table 5.3. **Development of area and livestock participation under the measures between 1993 and 2002**

	Units	1993	1994	1995	1996	1997	1998	1999	2000	2001	2002	2003
Total area[1]	ha	1 061 840	1 061 840	1 061 840	1 082 490	1 075 728	1 075 405	1 071 899	1 072 492	1 071 130	1 069 770	1 067 055
Total livestock units (LU)[2]	LSU	1 375 831	1 375 831	1 330 282	1 336 189	1 307 714	1 303 255	1 304 285	1 299 512	1 310 346	1 305 363	1 287 028
General direct payments												
Area payments (since 1999)	ha							1 021 945	1 029 899	1 028 877	1 023 819	1 027 321
Complementary direct payments (until 1998)	ha	1 020 858	1 001 300	957 014	968 545	971 233	976 422	0	0	0	0	0
Payments for integrated crop production (until 1998)[2]	ha	179 152	298 297	364 414	646 282	784 562	833 530	0	0	0	0	0
Payments for keeping grazing farm animals	Cows/LU	68 061	68 726	71 566	72 630	73 560	74 999	289 467	298 112	311 283	329 702	336 891
Payments for keeping livestock under difficult conditions	GVE	503 211	480 923	477 506	473 877	463 354	456 466	455 177	450 313	452 093	529 908	525 163
Payments for farming on steep slopes	ha	309 693	242 503	239 795	238 239	235 170	234 810	232 020	233 219	233 020	231 069	230 577
Ecologically motivated payments												
Payments for ecological compensation	ha	69 393	74 099	78 139	94 039	103 919	107 892	107 298	111 851	117 302	119 729	121 010
Payments for ecological quality	ha	0	0	0	0	0	0	0	0	0	15 552	26 921
Payments for green fallow[3]	ha	1 104	2 003	2 804	4 805	6 841	8 245	0	0	0	0	0
Payments for extensive cereal and rapeseed cultivation	ha	72 960	81 858	80 370	79 467	95 612	91 402	87 761	83 577	81 576	80 140	78 425
Payments for organic crop farming	ha	18 908	21 223	28 350	53 982	66 885	72 466	78 454	82 822	93 565	102 802	110 134
Payments for turning animals outdoors regulary	LU	91 412	117 952	146 283	254 759	355 513	434 550	538 667	618 000	690 939	742 993	793 517
Payments for animal housing systems	LU	0	0	0	94 145	139 707	171 462	225 434	265 236	310 139	345 763	384 969
Payments for summer pasturing	LU	302 403	301 416	310 184	315 632	310 965	306 203	297 015	312 477	308 418	291 610	315 156

1. Source: Statistical surveys and assessments; 1993 and 1994 no data (= Utilised Agricultural Area (UAA) for 1995).
2. Source: Statistical surveys and assessments; 1994 no data (= value as per 1993).
3. Expiring regulation of 1999-2000, as per 1999 the respective areas are no longer published.
Source: 1993 to 1998: Report on the disbursement of direct payments (BLW, various years), 1999 to 2003: Agricultural Report (BLW, various years).

Between 1993 and 2003, the direct payments disbursed have risen from CHF 1.28 billion to CHF 2.47 billion (see also Table 5.2). In 2003, general direct payments accounted for the largest share, namely 80.9%, of these payments. The financial importance of ecological and ethological contributions is relatively small (15.4%) by comparison. With the exception of the adaptation of payments for keeping livestock under difficult production conditions already mentioned above, general direct payments have not risen any further since 1999. On the other hand, ecological and ethological contributions still show a slight increase, which can be explained by the fact that there is a steady rise in the number of farms taking part in the programmes (Figure 5.5 and Table 5.3).

Figure 5.5 shows the development of the shares of area and LSUs which farms put into programmes for general direct payments. Figure 5.5 is based on the data presented in Table 5.3, whereby contributions for grazing livestock units replace those for farmers who keep cows but do not produce traded milk. Thus, a considerably larger number of animals (roughly 200 000 LSUs, see Table 5.3) is eligible for these payments. Furthermore, the increase in payments since 1999 may also be attributable to the rising number of withdrawals from milk production. This assumption is supported by the increasing number of LSUs participating as shown in Table 5.3.

Figure 5.5 shows clearly that in 2003, area payments were granted for 96% of the total agricultural land in Switzerland. At the same time this means that the criteria for proof of ecological performance are observed on these areas, i.e. the managers of these areas can provide this proof. In 2003, a total of 57 397 farms, or roughly 89% of all the farms in Switzerland, fulfilled the conditions for the receipt of area payments. As a result of the system itself, the shares held by other contributions is lower since, given the objectives foreseen under the programmes, neither every type of animal nor the entire agricultural area are eligible to receive payments.

Figure 5.6 depicts the development of ecological direct payments. The number of participants in the RAUS (outdoors) and BTS (particularly animal-friendly housing) programmes is still rising strongly and accounts for significant shares – 62% (RAUS) and 30% (BTS). The lower share held by the BTS programme results from the higher requirements on the buildings involved. Those farms that already have buildings which fulfil the programme's criteria (housing with outdoor yard) are eligible to participate, while the remaining farms are obliged to invest in their animal housing if they wish to take part.

Organically worked areas held a share of about 10% in 2003. Switzerland has a total of 65 866 farms of which 6 186 or 9.4 % are worked according to the principles of organic farming. This puts Switzerland in the lead compared to rest of Europe and is attributable largely to marked growth at the beginning of the 1990s.

On the other hand, ecological compensatory areas and extensive cereal cultivation hold a modest share. With regard to ecological compensatory areas, it must be mentioned that in order to qualify for direct payments farms must put a percentage of their area into the ecological compensation programme, namely 5% as per 1997, and 7% from 1998 onwards. Independently of these limits, contributions for services in the ecological compensation sector are reimbursed by specific payments. In 2003, payments for extensive meadowland and for high-standing fruit trees represented the most important elements of ecological compensation, accounting together for about 70% of the payments disbursed for ecological compensation.

Figure 5.5. **Development of the area and LSU shares
in the general direct payment programmes**

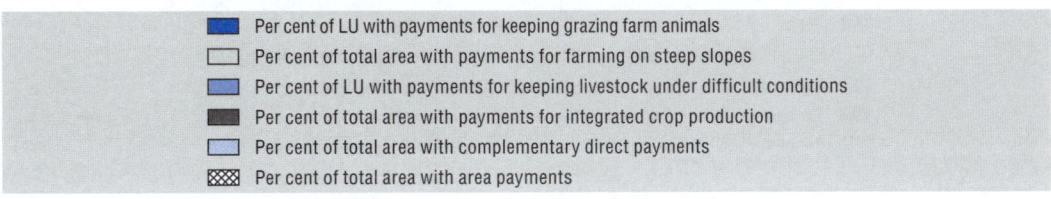

- Per cent of LU with payments for keeping grazing farm animals
- Per cent of total area with payments for farming on steep slopes
- Per cent of LU with payments for keeping livestock under difficult conditions
- Per cent of total area with payments for integrated crop production
- Per cent of total area with complementary direct payments
- Per cent of total area with area payments

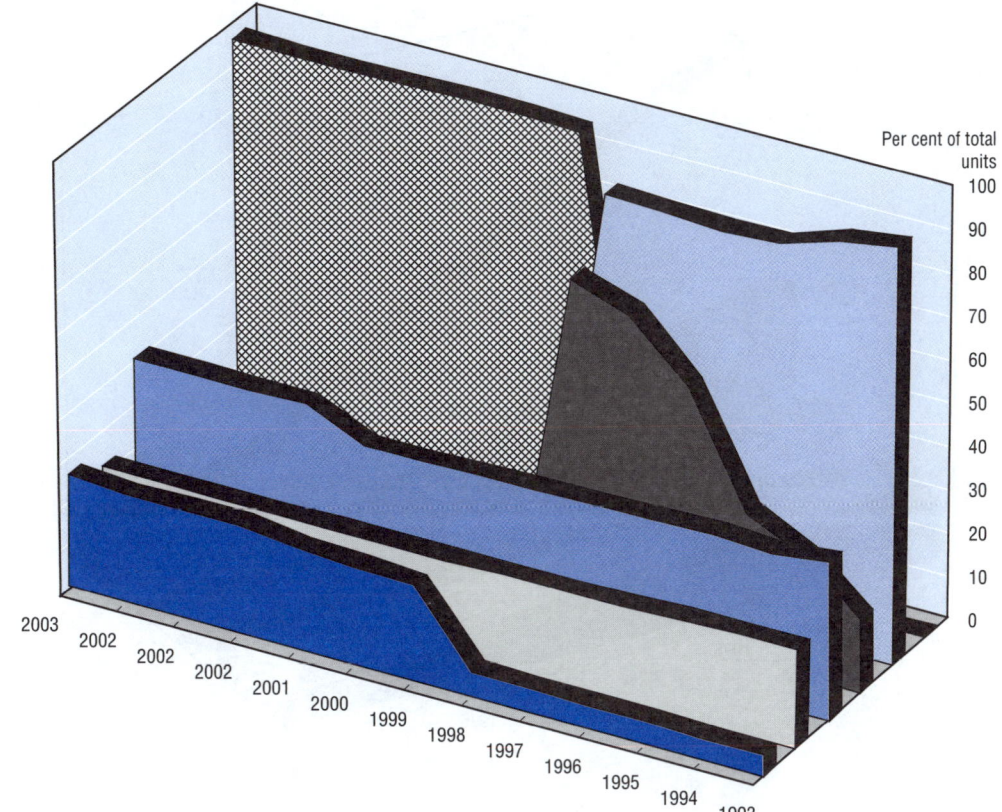

Source: Table 5.3.

Figures 5.5 and 5.6 illustrate clearly the differing effects over time of the Red and Green Ticket Approaches. While in most cases Red Ticket measures exhibit constant participation at a high level, Green Ticket measures are generally characterised by rising participation. After a certain time, this stabilises at a steady level, as it is not beneficial for those farms which already participate to commit more land or animals to the programme or to go in for the programme.

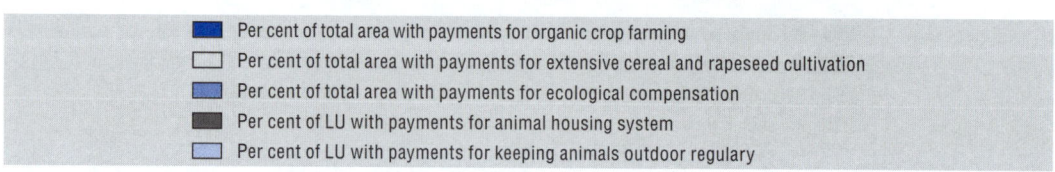

Figure 5.6. **Development of area and LSU shares in programmes for ecological and ethological direct payments**

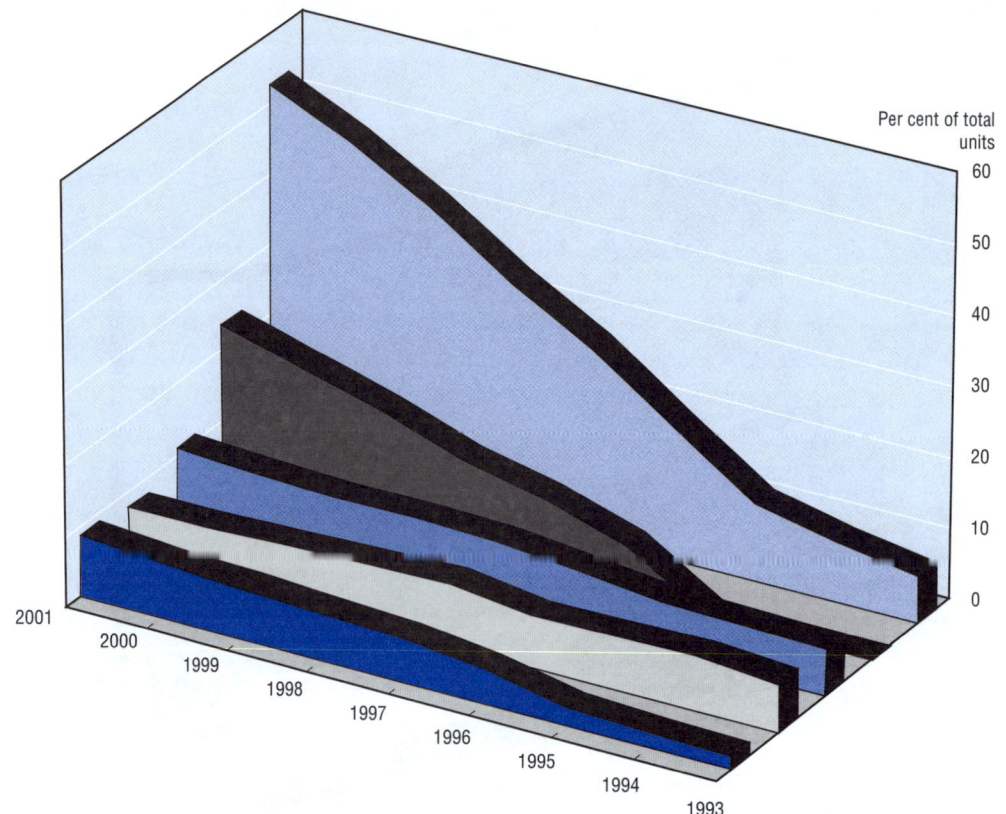

Source: Table 5.3.

5.3. Estimation of policy-related transaction costs

This section deals with the most important fundamentals which serve as a basis for the estimation of transaction costs. The general concept of the assessment is first presented. The organisational implementation of the Swiss direct payment system is then described and the two case study of Cantons, Grisons and Zurich are introduced. The last two sub-sections outline the actual procedure for the calculation of transaction costs at the various administrative levels and the key figures used for interpreting direct payments.

General concept for the calculation of policy related transaction costs

By way of an introduction to the origins of the calculation concept, Figure 5.7 depicts a general system for the implementation of political measures – in this case direct payments – with the actors and their interrelationships. Each actor receives or passes on

THE IMPLEMENTATION COSTS OF AGRICULTURAL POLICIES – ISBN 978-92-64-03091-6 – © OECD 2007

Figure 5.7. **Flowchart and processes in a general implementing and monitoring system**

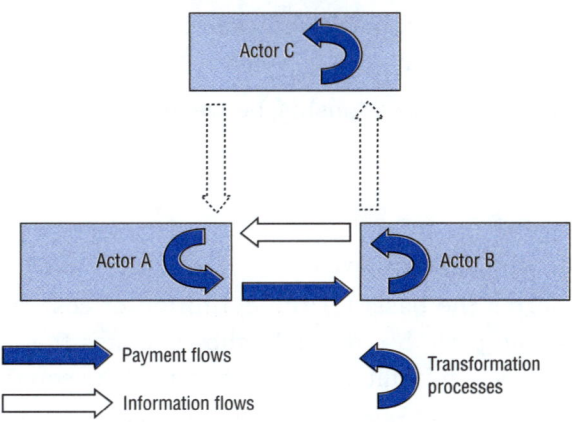

information or money. In order to perform these actions, the actors need an internal transformation process: information is received, processed and passed on, the flow of money or of payments is generated.

Each arrow and process illustrated here is relevant to the transaction costs, whereby there is a fundamental difference between transport costs and transformation costs. This is illustrated in Figure 5.8. In addition, we differentiate between transport costs for information flow and the flow of money (direct payments).

Figure 5.8. **PRTC in a general implementing and monitoring system**

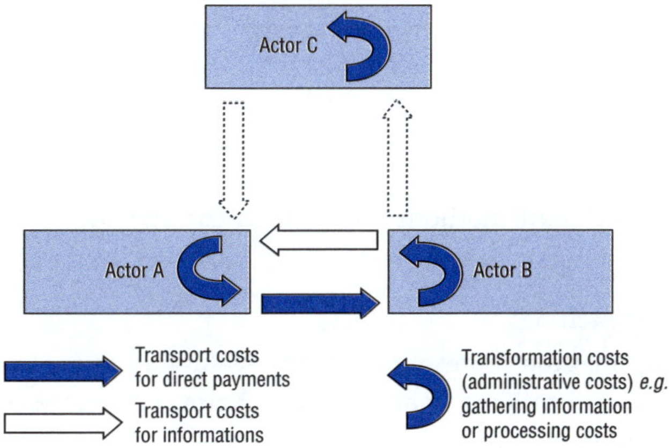

PRTCs (in or outflow) arise for each actor at each arrow; it is not apparent who pays the costs.

The multilevel measures of the direct payment system give a further dimension to the system shown in Figure 5.8, whereby interrelationships also exist between the processes and flows of the individual measures. For example, the control of the proof of ecological performance applies to several measures.

Each actor within the system generates costs; that is, both transformation costs and transport costs. The party who generates these costs must not necessarily be the one who has to pay for them. This applies in particular to private actors or corporate bodies who invoice their services (costs) to other actors (*e.g.* private control organisations).

Given this general knowledge of the system, the following basic information is essential for determining transaction costs:

- knowledge regarding the relevant participating actors;
- knowledge regarding the political measures;
- knowledge regarding the interrelationship between the actors in the context of the individual measures;
- knowledge regarding the transformation processes of the actors in the context of the individual measures; and
- knowledge regarding who pays the costs.

This knowledge forms the basis for the definition of cost centres which contain information regarding who pays the costs (T), who generates the costs, *i.e.* actor (V) and political measures (M), or which facilitates differentiation between these parties.

Thus, a specific cost centre K related to a measure can, in general, be referred to as K_{TVM}. The cost centre is defined by a function of the cost-generating process and structure characteristics.

K_{TVM} = f (processes, structure characteristics)

If all the cost centres can be obtained according to this principle, it is possible to tabulate the results (Table 5.4). The required results can be derived by simple additions *via* the respective indices.

Table 5.4. **Illustration of the general concept of PRTC acquisition by cost centres**

Measure M	Payer 1	Payer 2	
	Actor 1	Actor 2	Actor 3
M1			K_{T2V3M1}
M2			
M3			

There are two different methods with which the specific cost centres K_{TVM} can be determined.

Bottom-up or input method:

The bottom-up method represents a direct evaluation of the cost function, whereby every structure and process characteristic (factors) generates costs. The respective costs are attributed to each factor.

K_{TVM} = f(F)

f(F) = Sum($a_i * F_i$) whereby a_i = costs and F_i = factors

The use of the bottom-up method on its own is only possible for cost centres for which all the cost-generating factors (inputs) and their costs are known. However, this is not always the case. This problem does not arise with the top-down method.

Top-down or output method:

In the case of the top-down approach, the transaction costs of a certain measure are calculated on the basis of the overall costs of a payer. It is often easier to determine a sum for cost centres than the costs of an individual measure. For example, a payer's overall costs may be known through the budget position of a public institution, while their distribution between cost centres is not available. The top-down approach deals with this distribution by dividing a fixed output, such as the State's total implementation costs, between the cost centres. However, this demands precise knowledge of the processes which generate this output.

The first step involves the allocation of an institution's overall costs to the various processes. In the second phase, assumptions can be made regarding the distribution of these partial costs between the cost centres (Table 5.5).

Table 5.5. **Procedure for the top-down method**

1st Step	
$V_1 = C$	C = absolute sum of transaction costs of actor V_1
$V_1 = \text{Sum}(a_j * V_1)$	with $\text{sum}(a_j) = 1$ and $a_j * V_1$ = costs for process P_i
	a_j = evaluation of partial expenditure/partial costs for process P_i
2nd Step	
$P_i = \text{Sum}(a_{V1Mi} * P_i)$	with $\text{sum}(a_{V1Mi}) = 1$ and $\text{sum}(a_{V1Mi} * P_i) = a_j * V_1$ = costs for process P_i
	a_{V1Mi} = evaluation

For example, the overall costs for implementation in Canton Grisons are known. A part of these costs are generated by the number of farms, regardless of their farming practices and participation in measures. The evaluation of this share corresponds to the evaluation of a_i in Step 1. The value a_{V1Mi} must be evaluated in Step 2 in order to allocate these partial costs P_i to the measures. Two variants for this evaluation are presented later in this section.

The values for K_{TVM} can be evaluated with statistical methods using a time series analysis for one payer or a cross-section analysis involving similar payers (*e.g.* Huber, 1998; or Mann, 2001). Good values for C and sufficient data points are both essential for the calculation of useful results. To a large extent, the relevant bases for calculation for the Swiss direct payment system are lacking or can only be acquired at great cost.

As opposed to statistical evaluations, the allocation of costs to the various cost centres can also be realised on the basis of assumptions. Detailed knowledge of the implementation and control processes is a prerequisite for fixing these assumptions. Furthermore, the effects of the assumptions reached can be tested and the results delimited within a reasonable scope by means of variant calculations (for both methods).

Organisation of the implementation of the Swiss direct payment system (processes)

The implementation of the Swiss direct payment system is subject to the terms of the federal direct payment regulation DZV (SR 910.13). The actors and their duties are described in Section 4 of this regulation, as illustrated in the columns (Figure 5.9). The lines show the principal elements of implementation, while the arrows indicate the relationship between the actors. These combine in the white fields to show the actors' main processes. The costs of these processes flow into the evaluation of the transaction costs. These processes are described in detail in Section 5.4 which presents the procedure for the evaluation of transaction costs at the various administrative levels.

Basically, the implementation of the direct payment system involves three groups of actors: farmers or farms, the cantons and the state.

Figure 5.9. **Actors and processes in the implementation of the Swiss direct payment system**

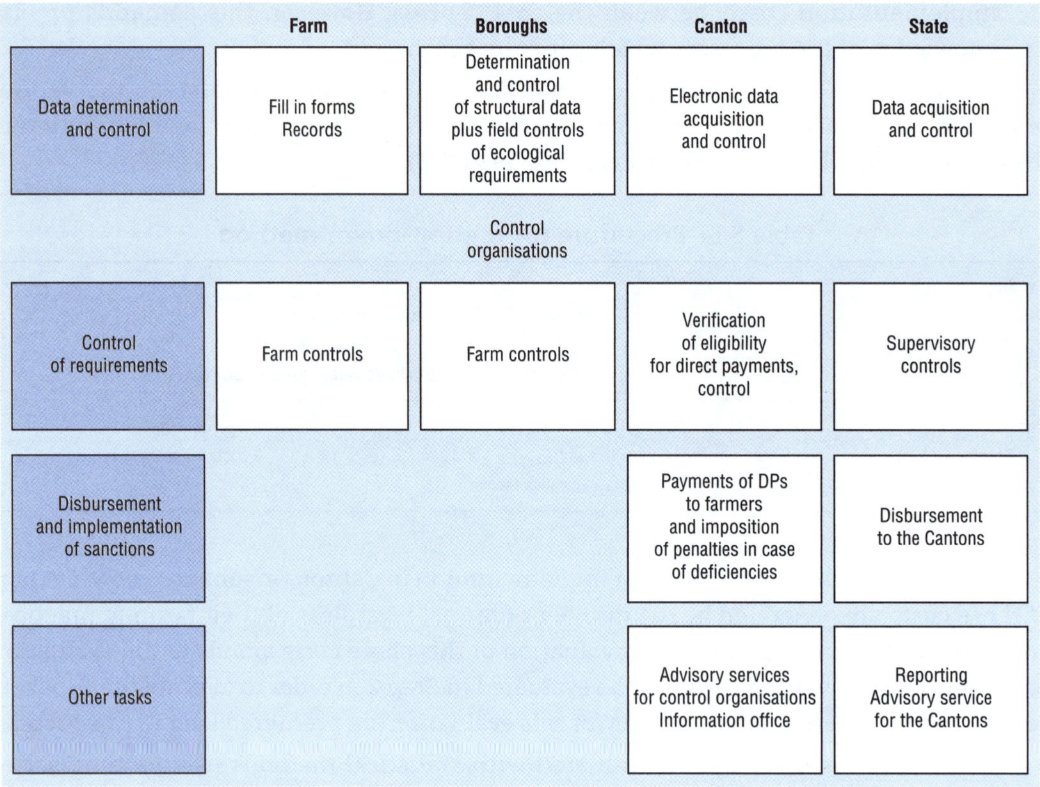

	Farm	Boroughs	Canton	State
Data determination and control	Fill in forms Records	Determination and control of structural data plus field controls of ecological requirements	Electronic data acquisition and control	Data acquisition and control
		Control organisations		
Control of requirements	Farm controls	Farm controls	Verification of eligibility for direct payments, control	Supervisory controls
Disbursement and implementation of sanctions			Payments of DPs to farmers and imposition of penalties in case of deficiencies	Disbursement to the Cantons
Other tasks			Advisory services for control organisations Information office	Reporting Advisory service for the Cantons

Farmers are integrated into the implementation process in three ways: they must complete application forms for the respective direct payment; they must keep appropriate records of their activities on the farm over the year to provide a basis for the controls and they play an active role when the controls take place on the farm.

The cantons are the most important executive institution. They organise data determination, on-farm controls requirements, verify eligibility to receive direct payments, impose penalties in case of deficiencies, and disburse the direct payments to the farmers. At the same time, they are a link in the chain of communication between legislators and the individual providing the service (farmer). The cantons are free to out-source part-sectors, such as controls or data determination, to other actors. In most cantons, the boroughs are involved in the fields of data determination and imposing ecological requirements (*e.g.* control of due date for mowing ecologically compensatory areas). In some cantons (*e.g.* Zurich), on-farm controls (fulfilment of proof of ecological performance, RAUS, BTS) are carried out by private organisations. Control of organic farming is out-sourced to private organisations in all cantons.

The Ministry of Agriculture (BLW) is the highest instance and as such has the task of supporting and controlling implementation at the cantonal level. In addition, the Ministry disburses the contributions to the cantons and reports on implementation.

Case studies

The description of the relevant processes reveals that the cantons play a leading role in the implementation and control system. to a large extent, cantons have the authority to decide for themselves on the manner in which they perform their duties. Accordingly, different patterns have developed with regard to implementation. This applies in particular to the degree of out-sourcing of the control organisations. In addition, there are minor differences in the organisation of data determination. Due to the differing systems employed by the Cantons, the transaction costs for the whole of Switzerland can only be obtained if every Canton is included in the investigation.

In this case study, the investigation is limited to two case studies of the cantons Grisons and Zurich. The control systems utilised by theses cantons differ in that in Grisons the canton runs the control office, while in Zurich it is completely out-sourced to the private organisation Agrocontrol. In addition to these organisational differences, the two cantons are ideal for the case study as they are relatively large and their farming practices cover most types of farm and agricultural zones. Table 5.6 presents a comparison of the Cantons Zurich and Grisons based on organisational and structural criteria.

Table 5.6. **Organisational and structural differences of the case study cantons**

	Canton Grisons	Canton Zurich
Organisation		
Control office	Cantonal control office	Agrocontrol (Farmers' Union)
Data acquisition and field controls	Boroughs	Boroughs
Electronic data acquisition	Canton	By students
Structural characteristics		
Number of boroughs with agriculture	207	169
Number of farms with direct payments	2 745	3 657
% organic farms	50.1%	12.9%
% farms in mountainous area	92.4%	11.4%
% organic farms in mountainous area	53.9%	23.6%
Average size of farms eligible to receive direct payments	19.13 ha	19.41 ha
Total direct payments	167 Mio. CHF	141 Mio. CHF
Direct payments per farm	60 838 CHF	38 556 CHF
Direct payment/ha exploitable surface area	3 180 CHF	1 986 CHF

Source: Own surveys plus AGIS-Data Bank (BLW, 2003a).

Overall, while Canton Grisons has fewer farms, the average size of the farms which are eligible to receive direct payments is practically identical. In Canton Zurich there are far less organic farms and farms situated in mountainous areas. The share of organic farms is larger in mountainous areas, in particular in Canton Zurich. Roughly CHF 167 million are disbursed in the form of direct payments in Canton Grisons, not counting summer pasture contributions and contributions under the terms of the ecological quality regulation. Canton Zurich receives about CHF 140 million. Thus, per farm and per hectare agricultural land, Canton Zurich only accounts for a little over 60% of the payments received by Canton Grisons. This is mainly due to the lower arable share, the higher rates for mountainous areas and the lower importance of dairy farming in Canton Grisons.

In order to evaluate the transaction costs for the case studies, it is essential that the transaction costs at farm level of all the farms within both cantons are included. The same applies to the costs incurred by the boroughs and cantons. On the other hand, at the State level, the transaction costs for the chosen cantons must be isolated from the total expenditure on the implementation of direct payments.

Procedure for calculating policy-related transaction costs

General summary of procedure

In this case study, the estimation of transaction costs and their allocation to the individual measures differ for each actor. However, a common element is that every farm within a canton is assigned a cost share pertaining to each actor. The values K_{TVM} are calculated for each farm and are then added again to compile the transaction costs per actor.

Depending on the data available, an actor's overall or partial costs are calculated using either the top-down or bottom-up procedure. Every actor exhibits cost items which even experts find difficult to allocate directly to a specific measure. In this case, Step 2 of the top-down procedure is applied, whereby two different methods are used to determine the values a_{VM}.

Variant "Participation":

In the case of the "Participation" variant, a farm's participation or non-participation in a measure is the only relevant factor for allocation to the measures. For example, if a farm takes part in all the measures, the costs of completing the direct payment form are distributed evenly over all the measures.

> $a_{VM} = 1/\text{sum}(M) * M$; whereby M = 1 if a farm participates in measure M and M = 0 if it does not participate in measure M.

Variant "Direct payment share":

> On the other hand, in the case of the "Direct payment share" variant, the costs are distributed over a farm according to the direct payment shares of the measures.
>
> $a_{VM} = \text{direct payments (M)}/\text{sum(direct payments)}$ of the individual farm

The following sections are devoted to describing the procedure for calculating and allocating costs to the measures for all the actors involved. In conclusion, the process is summarised formally in Table 5.7.

Procedure at farm level

Figure 5.10 illustrates the processes which are relevant for calculating transaction costs at farm level, whereby labour costs are the main element. Labour expenditure relating to the individual processes is estimated by means of expert interviews and multiplied by pre-set labour costs. The influence of uncertainties in these estimations is tested by means of variant calculations using different labour costs (CHF 0 to 25 per hour).

The costs are calculated for each individual farm. In the case of certain items, a farm's participation/non-participation in a measure (*e.g.* record of time outdoors for the RAUS programme) is decisive. If this applies, the respective cost share is allocated directly to a measure. On the other hand, in the case of other items, it is only relevant if a farm receives direct payments or if the costs are related to structural characteristics of the farm. These costs are allocated to the measures according to the variants mentioned earlier in this section.

Table 5.7. **Procedure for cost allocation to the individual farms and measures**

Level	Actor	Cost factors in CHF	Distribution key – $K_{Mi} = a_{Vmi} *$ cost factor	Explication for the cost assignment to the individual farm
State	State	Cost factor Canton	/K/BK	Allocation *via* number of Cantons and number of farms per Canton
		Cost factor farm	/BET	Allocation *via* number of farms in CH
		Cost factor measure	/MET/a_{Vmi}	Allocation *via* number of farms participating in CH
		Cost factor implementation quality	$*A_i$/BK	Allocation *via* quality characteristics of the Canton and number of farms in the Canton
Canton	Canton	Cost factor farm	/BET	Allocation *via* number of farms in the Canton
		Cost factor measure	/MET/a_{Vmi}	Allocation *via* participating farms in the Canton
		Cost factor not assignable	/BET	Allocation *via* number of farms in the Canton
Borough	Borough	Fixed costs borough	/BET	Allocation *via* number of farms in the borough
		Cost estimate per farm		No allocation necessary
		Cost factor measure	/MET*/a_{Vmi}	Allocation *via* participating farms within the borough
Canton	Control organisation PEP	Cost factor farm		Fixed lump sum per direct payment farm (BP)
		Cost factor measure		Fixed rate per participation in measure (A)
Farm	Bio Inspecta	Cost factor farm – BP	/a_{Vmi} *BIO for i = organic farming with condition $K_{Mi} >= 0$	No allocation necessary
		Cost factor measure – A	/a_{Vmi} *BIO for i = organic farming with condition $K_{Mi} >= 0$	No allocation necessary
	Farm	Cost factor farm	/a_{Vmi}	No allocation necessary
		Cost factor measure		No allocation necessary
PRTC of a measure Mi in a case study Canton			Column sum *via* farms	
PRTC of a case study Canton			Column sum *via* farms and measures	

A_i = Distribution key – implementation quality for Canton i
a_{Vmi} = Key for allocation to measure Mi
BA = Type of farm – BIO or PEP with BIO + PEP = 1
BET = Sum of the farms at a level (first column of Table 5.7)
BK = Sum of the farms in the case study Cantons
K = Sum of all Cantons
KM_i = Partial costs of an actor for measure M_i (partial costs of K_{TVM})
MET = Sum of the farms at a level, which participate in measure Mi
ÖLNET = Number of PEP farms at a level (first column)

Figure 5.10. **Processes taken into consideration at farm level**

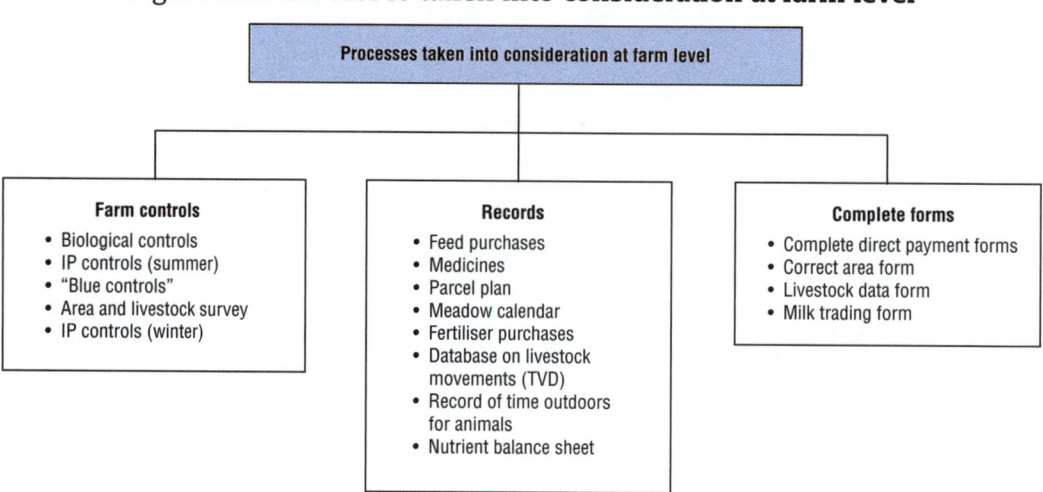

Procedure at control organisation level

Control organisations taken into account are presented in Figure 5.11. The costs arising from the respective controls are assigned to each individual farm on the basis of its structural data. In the case of organic farms, the rates of the control organisation for organic farms (Bio Inspecta) are applicable. Costs for any subsequent checks or reductions (*e.g.* loyalty bonus) are not taken into consideration as these are attributable to the farm management and not to the direct payment system. In addition, costs and controls associated with the certification of products are not taken into account (*e.g.* wine cellar controls).

In Canton Zurich, the rates of the control organisation (Agrocontrol) are applicable for both PEP farms (proof of ecological performance) and organic farms. The details provided by Agrocontrol (overall costs) serve as the test value. In Canton Grisons, the overall costs of the control organisation are known. As is customary, each PEP farm is charged a lump sum.

Figure 5.11. **Control organisations taken into consideration**

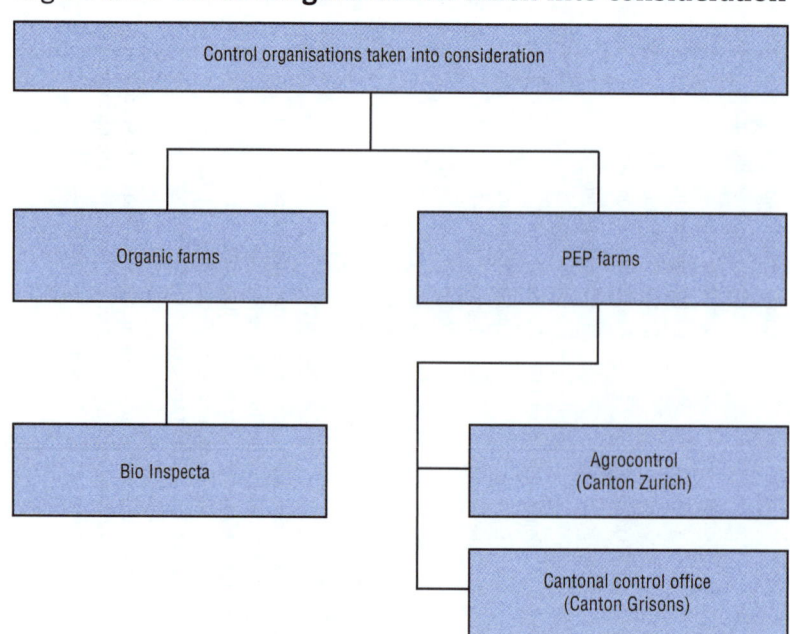

In addition, the following assumption is reached in relation to the allocation of control costs to measures: the difference between organic controls and PEP controls is assigned to the measures for organic farming. The remaining costs are distributed according to the variants described earlier in this section.

Procedure at the borough level

Basically, the tasks of the boroughs in the cantons covered by this investigation do not differ. However, while these tasks are set down in a duty roster in Canton Zurich, the boroughs in Canton Grisons are free to carry out these duties at their own discretion. As a result, there are no uniform implementation processes in Canton Grisons. Therefore, two different variants are considered in Canton Grisons. The processes are assigned to labour expenditure and costs on the basis of surveys. Finally, the more expensive, but more widespread variant was chosen.

Canton Zurich has an expenditure estimate for arable land sites. An average estimate of 3 hours per farm and year is assumed for each farm. Costs for further training of the site manager and expenses are added to this. These are estimated for the calculation of the transaction costs.

If a direct assignment of partial costs is not possible, both Cantons carry out allocation to measures in accordance with the two recognised variants presented in this section.

Procedure at the cantonal level

The costs of the cantons are obtained by means of questionnaires covering the following aspects:

- registration of the departments and persons involved, whereby the labour expended by all staff for the implementation of direct payments is recorded;
- registration of the gross wage costs of the individual employees, the department's infrastructure costs and the costs of purchased services;
- determination of the influence values on the labour expenditure (cost factors).

Some of these costs are assigned directly to the measures on the basis of the information obtained from the survey (Figure 5.12). In addition, there are cost items which can be attributed to the number of farms, or which can be regarded as fixed costs. These latter are not connected to the number of farms or the farms' participation in direct payment programmes.

Costs which cannot be assigned directly and the costs of the cost factor "number of farms" are allocated according to two recognised methods.

Procedure at the state level

Cost determination at State level is carried out in the same way as for the cantons, namely by means of a questionnaire. Labour expenditure of the staff involved in implementation and the associated costs, infrastructure costs and the costs of purchased services are likewise determined at State level. The factors which define the total cost expenditure (Figure 5.13) differ slightly from those which apply to the cantons. In this case too, there are cost shares which can be attributed directly to a farm's participation in certain measures and costs which must be allocated.

The number of farms and the number of cantons in Switzerland are responsible for a part of the costs. The quality of implementation as performed by the Cantons is a further cost factor. A canton with high quality implementation generates fewer costs at state level

Figure 5.12. **Cost factors at the cantonal level**

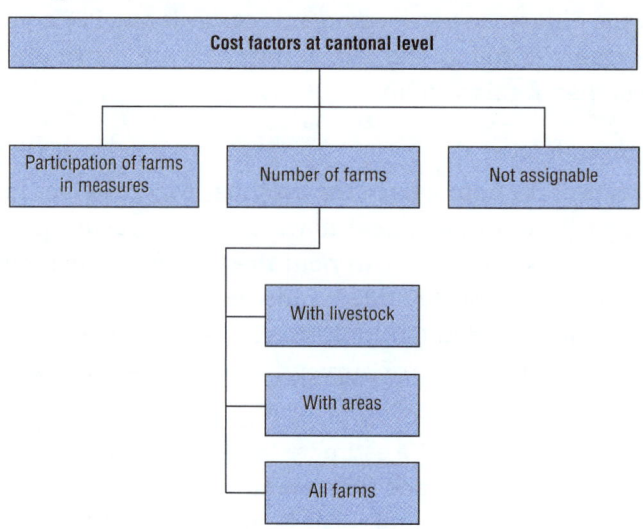

Figure 5.13. **Cost factors at state level**

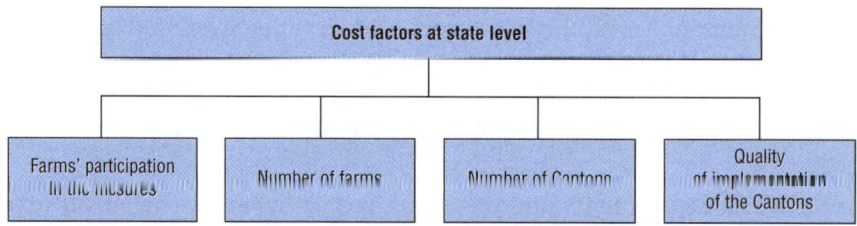

than a canton with a lower quality implementation. The following procedure is used to determine the relevant cost shares of the cantons in the State's costs:

● Cost shares which can be attributed directly to measures are assigned directly to the farms and measures via the participation of the farms.

● In the case of the cost factor "number of farms", the Canton's cost share is calculated on the basis of the number of farms (number of farms Canton/number of farms Switzerland).

● The same cost share is assigned to all Cantons on the basis of the cost factor "number of Cantons".

A canton's share in the cost factor "quality of implementation of the cantons" is determined by the share of the farms queried in relation to the number of farms eligible for direct payments. This procedure is based on the assumption that, geographically speaking, the share of incorrectly run farms in Switzerland is evenly distributed. If the number of farms queried by a canton exceeds the Swiss average, this is regarded as evidence of high quality implementation. Consequently, this canton is expected to generate fewer costs for the state than a canton with a lower number of queried farms.

The cantons' shares in the costs at state level which are determined in this way are allocated to the measures (in the same way as the costs of the cantons).

THE IMPLEMENTATION COSTS OF AGRICULTURAL POLICIES – ISBN 978-92-64-03091-6 – © OECD 2007

Overview of the procedure

Table 5.7 presents a formal overview of the chosen procedure for all actors.

Example: A part of the costs incurred by the boroughs are fixed costs (cost factor "fixed costs borough"). A part of these costs is allocated to each farm (cost factor "borough/BET"). In this case, BET is the number of farms in a borough. This sum is assigned to the measures M_i with the factors a_{Vmi}. The result of a_{VMi} depends on the variants presented in the last sub-section of Section 5.4.

This results in the following function for cost centre K_{Mi} for the borough's fixed costs for every farm: $K_{Mi} = a_{VMi} *$ cost factor borough/BET.

Key figures of policy-related transaction costs

Transaction costs are presented as key figures in order to facilitate comparison with other cantons or investigations. Three key figures are used:

- PRTC per CHF of direct payment as indicator for the efficiency of fund transfer.
- PRTC per relevant unit (per hectare or LSU) to describe the cost function dependent on the factors land, land utilisation and number of animals.
- PRTC per farm as a measure for the average total costs of a participating farm.

5.4. Results of estimations in Cantons Grisons and Zurich

In this section, the estimated transaction costs of the Swiss direct payment system for the Cantons Grisons and Zurich are identified and interpreted. The influence of assumptions and variants on the transaction costs is also discussed. The results of both case studies are presented according to the same pattern:

- Discussion of the basic variant: the transaction costs are allocated to the different cost centres on the basis of the variant "Participation". On-farm labour costs are calculated at CHF 20 per hour.
- Assessment of the influence on transaction costs of the assumptions reached *via* the variant calculations.

Then follow the analysis and discussion of the cost differences exhibited by the two cantons and the structural and organisational factors which are relevant to the costs.

Policy-related transaction costs in Canton Grisons

Transaction costs in the basic variant (Canton Grisons)

Table 5.8 shows the transaction costs of direct payments for Canton Grisons according to the basic variant. The costs of implementation and controls relating to general direct payments are shown in the upper section, while those relating to ecological direct payments and ethological contributions are presented in the lower section. The transaction costs are allocated not only to the parties who generate them, but also to the measures and payers. At farm level, the costs are also subdivided into the sectors controls, records and forms. When considering the payer, it must be borne in mind that up until 2002, Canton Grisons subsidised a part of the costs of the control organisation for organic farms and the cantonal control office (an overall total of CHF 160 000 per year). These subsidies are not taken into account in Table 5.8. In addition, the total costs of the

Table 5.8. Transaction costs in Canton Grisons (basic variant)

CHF

Payer	State	Canton	Boroughs	Farm					Total	Share paid by authorities
Actor	State	Canton	Boroughs	Control organisations for organic farms	Control organisations for PEP farms	Records	Forms	Farm controls	Total	Share paid by authorities
Policy										
Area payments	9 156	104 887	45 403	29 180	29 096	138 677	26 059	14 575	393 032	159 445
Payments for keeping grazing farm animals	8 597	95 893	41 582	27 711	24 886	127 291	23 721	13 268	359 950	146 073
Payments for keeping livestock under difficult conditions	8 666	95 895	41 581	27 756	24 846	128 838	23 721	13 268	361 572	146 142
Payments for farming on steep slopes	8 546	94 577	40 951	27 556	24 454	124 445	23 423	13 101	354 052	144 073
Payments for ecological compensation	9 012	113 789	44 610	27 960	28 316	139 634	25 569	14 301	400 190	167 411
Payments for extensive cereal and rapeseed cultivation	911	19 960	4 128	7 961	2 714	19 654	2 277	1 273	52 877	24 998
Payments for organic crop farming	5 951	71 610	19 382	35 904	0	51 067	11 132	58 970	574 017	96 944
Payments for animal housing systems	2 311	40 288	10 194	6 647	3 593	33 061	5 728	6 939	110 761	52 793
Payments for turning animals outdoors regulary	7 411	102 470	35 789	27 850	18 150	153 696	20 326	31 144	393 835	145 669
Total Actor	60 560	739 369	283 622	51 524	156 055	916 363	161 955	166 839	3 000 287	1 083 550
Total Payer	60 560	739 369	283 622	191 736						

Table 5.9. Key figures of transaction costs in Canton Grisons (basic variant)

CHF

	Total PRTC	Sum of payments	PRTC per payment (%)	Units		PRTC per unit	Sum of farms	PRTC per farm
Policy								
Area payments	393 032	62 736 704	0.6	52 299	ha	7.52	2 740	143.44
Payments for keeping grazing farm animals	359 950	29 834 828	1.2	36 445	LSU	9.88	2 631	136.81
Payments for keeping livestock under difficult conditions	361 572	38 667 105	0.9	40 254	ha	8.98	2 635	137.22
Payments for farming on steep slopes	354 052	14 041 723	2.5	32 079	ha	11.04	2 587	136.86
Payments for ecological compensation	400 190	5 925 862	6.7	14 556	ha	27.49	2 711	147.62
Payments for extensive cereal and rapeseed cultivation	52 877	319 048	16.5	798	ha	66.26	271	195.12
Payments for organic crop farming	574 017	5 893 005	9.7	28 617	ha	20.06	1 388	413.56
Payments for animal housing systems	110 761	1 330 812	8.3	13 765	LSU	8.05	746	148.47
Payments for turning animals outdoors regulary	393 835	7 839 470	5.0	43 702	LSU	9.01	2 350	167.59
Total	3 000 287	166 588 557	1.8	52 509	ha	57.14	2 745	1 093.00

public actors are listed in the last column of the table (sum of the state, canton and borough levels). The following findings can be derived from the composition of the transaction costs for Canton Grisons:

- The overall costs of implementing the direct payment system amount to roughly CHF 3.0 million. Public authorities pay about one third of the transaction costs, the farmers pay the remainder.

- About two thirds of the transaction costs paid by public authorities devolve upon the cantons, while the boroughs are liable for 26% of this sum. On the other hand, the share paid by the state amounts to just about 5%.

- Organic farming contributions account for the largest share of the transaction costs, namely about 20%, whereby the controls carried out by Bio Inspecta generate roughly 60% of this amount.

- The records kept by the farms generate just about one third of the total transaction costs, whereby BTS measures, Extenso contributions and organic farming occasion considerably less costs compared to the other measures.

The values for key figures of transaction costs can be calculated on the basis of Table 5.8. The results of the key figures are shown in Table 5.9:

- The overall share of transaction costs in the direct payments disbursed amounts to 1.8%. According to the basic variant, area payments exhibit the greatest efficiency with regard to fund transfer, followed by other Red Ticket measures, TEP (Payments for keeping livestock under difficult conditions), LSU contributions (Payments for keeping grazing farm animals) and slope payments. From this point of view, Extenso contributions (Payments for extensive cereal and rapeseed cultivation) and payments for organic production do considerably less well.

- The key figure of transaction costs per relevant unit reveals the sum of the implementation and control costs generated by a unit when participating in a measure. In relation to the total area involved, transaction costs of about CHF 57 per hectare occur in Canton Grisons. From this point of view, the area-specific Green Ticket measures appear to be of great value, while Red Ticket measures and ethological contributions are less useful. This is mainly due to the fact that the absolute costs of the measures are divided practically equally. Since a considerably higher number of farms take part in Red Ticket measures and ethological programmes, these incur less costs per unit than measures with a lower participation.

- If the transaction costs of a measure are allocated to the farms involved, it can be noticed straight away that organic farming contributions occasion high costs. This is due to the higher control costs paid by the farms. As a whole, these costs amount to an average of roughly CHF 1 093 per farm and year.

Influence of the choice of method and assumptions on transaction costs (Canton Grisons)

When interpreting and allocating transaction costs in the basic variant, the question arises regarding the influence on the results of the choice of method and the assumed labour costs or the estimated labour expenditure. In order to answer this question, Table 5.10 shows the key figures of the basic variant with different distribution ratios and labour costs.

Table 5.10. **Influence of the choice of method and labour costs on the key figures of the transaction costs in Canton Grisons**

CHF

Policy	Units	Baserun			Scenario								
					Influence of labour costs				Share of direct payments			
		PRTC per payment (%)	PRTC per unit	PRTC per farm	PRTC per payment (%)	PRTC per unit	PRTC per farm		PRTC per payment (%)	PRTC per unit	PRTC per farm		
Area payments	ha	0.63	7.52	143.14	0.01	0.16	3.05		0.86	10.35	197.48		
Payments for keeping grazing farm animals	GVE	1.21	9.88	136.31	0.03	0.21	2.91		0.21	1.69	23.39		
Payments for keeping livestock under difficult conditions	ha	0.94	8.98	137.22	0.02	0.19	2.93		0.52	4.95	75.59		
Payments for farming on steep slopes	ha	2.52	11.04	136.36	0.05	0.23	2.90		−1.10	−4.81	−59.65		
Payments for ecological compensation	ha	6.75	27.49	147.62	0.14	0.58	3.09		−4.90	−19.95	−107.12		
Payments for extensive cereal and rapeseed cultivation	ha	16.57	66.26	195.12	0.34	1.36	4.01		−11.11	−44.41	−130.78		
Payments for organic crop farming	ha	9.74	20.06	413.56	0.10	0.20	4.21		−0.97	−1.99	−41.09		
Payments for animal housing systems	GVE	8.32	8.05	148.47	0.16	0.15	2.84		−5.42	−5.24	−96.72		
Payments for turning animals outdoors regulary	GVE	5.02	9.01	167.59	0.12	0.22	4.16		−2.46	−4.40	−81.90		
Total	ha	1.80	57.14	1 109.0	0.04	1.11	21.29		0.00	0.00	0.00		

In Table 5.10 the influence of labour costs is shown as the range by which the key figures change when the assumed costs are raised or lowered by one franc. At the same time, these changes correspond to a 5% variation in the labour expended on the farms, for all processes, while labour costs remain unchanged. Table 5.10 shows quite clearly that the choice of the farms' labour costs or expenditure only has a slight influence on the key figures: a 5% increase in labour expenditure or a pay rise of CHF 1 per hour result in an overall increase of CHF 1.11 in total transaction costs per hectare of agricultural land. On the whole, the assumed labour costs have a much stronger influence on the PRTC of Green Ticket measures. In particular, an increase in labour costs leads to above-average changes in transaction costs in the case of contributions for ecological compensation (CHF 0.58) and extensive cultivation (CHF 1.36).

The results relating to the transaction costs per CHF 1 of direct payments (PRTC per payment) and to the transaction costs per farm (PRTC per farm) show that a change in labour costs only has a slight influence on the key figures. The rise in labour costs discussed above leads to an increase in costs of CHF 21 per farm or 0.04% for each franc disbursed for direct payments.

If those cost items of the measures which cannot be allocated are assigned by means of the variant "Direct payment shares", the key figures change in accordance with the third upper column of Table 5.10. However, since in this case it is only the allocation which is altered, the key figures of the respective measure change but not the results relating to the total transaction costs (bottom line of the table).

In the first instance, allocation of costs by means of the variant "Direct payment shares" raises the transaction costs of those measures with high direct payment shares (area payments, RGVE and TEP contributions). On the other hand, transaction costs sink in the case of measures with a lower share of the direct payments disbursed. By and large, the key figures are distributed evenly with this allocation. However, there is little change regarding the measures associated with organic farming contributions as the costs of these measures can, to a large extent, be allocated directly. Changes in transaction costs are illustrated again in Figure 5.14, whereby deviations of the costs per relevant unit compared to the basic variant are indicated.

Complementary to the key figures, the shares of the individual measures in the total transaction costs are illustrated in Figure 5.15, whereby the darker lines represent the basic variant, the lighter lines the "Direct payment shares" variant. For each, one variant is presented without labour costs and one with CHF 25 labour costs.

The path of the graphs shows that the share of organic farming contributions in the total PRTC is the only one to be influenced significantly by labour costs. The higher they are, the lower this share. This is mainly due to the relatively high fixed cost share of the organic farming controls. On the other hand, the choice of method has a marked effect on the distribution of the transaction costs. If the costs which cannot be allocated directly are divided according to direct payment shares ("Direct payment shares" variant) a considerably higher share is attributable to the measures associated with area payments, TEP and RGVE contributions. The opposite applies to the situation related to slope payments, contributions for ecological compensation, BTS and RAUS which have low shares in total direct payments.

Figure 5.14. **Influence of the variants on the key figures PRTC per relevant unit (Canton Grison)**

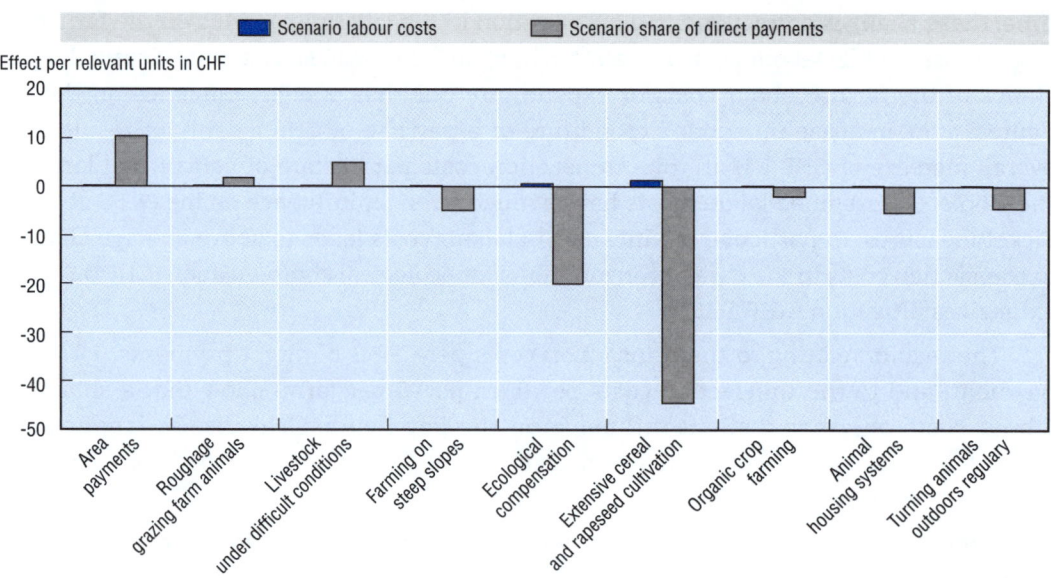

Figure 5.15. **Influence of the variants on the distribution of the PRTC according to measures (Canton Grisons)**

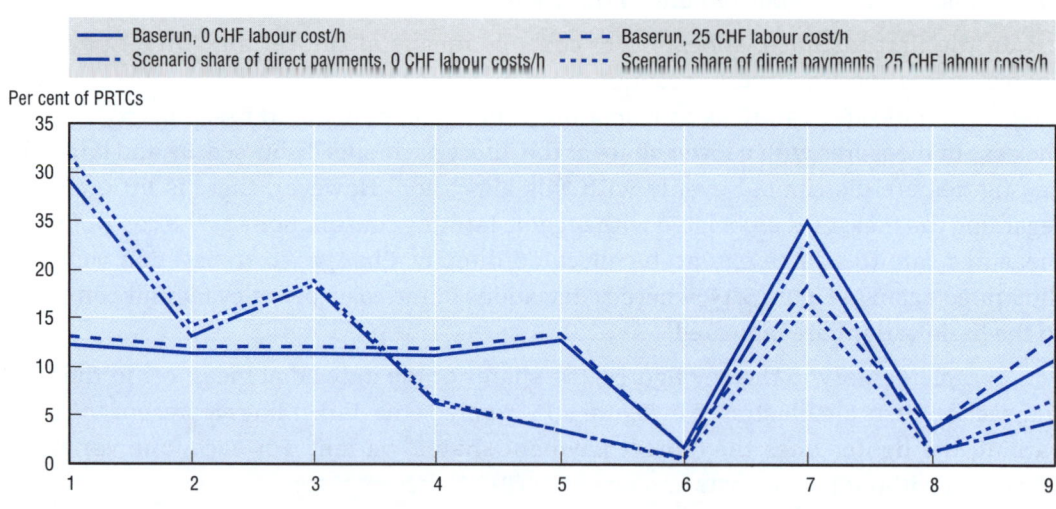

1. Area payments.
2. Payments for keeping grazing farm animals.
3. Payments for keeping livestock under difficult conditions.
4. Payments for farming on steep slopes.
5. Payments for ecological compensation.
6. Payments for extensive cereal and rapeseed cultivation.
7. Payments for organic crop farming.
8. Payments for animal housing systems.
9. Payments for turning animals outdoors regularly.

THE IMPLEMENTATION COSTS OF AGRICULTURAL POLICIES – ISBN 978-92-64-03091-6 – © OECD 2007

Policy-related transaction costs in Canton Zurich

Transaction costs in the basic variant (Canton Zurich)

Table 5.11 shows the transaction costs generated for Canton Zurich by the direct payment system in relation to actors, measures and payers when the basic variant is applied. The structure of the table corresponds to Table 5.8 for the case study Grisons. When reviewing payers in Canton Zurich, it must be borne in mind that, as in the case of Canton Grisons, the Canton subsidises the costs of organic farming and proof of ecological performance controls with a total of about CHF 220 000 per annum. These subsidies are not taken into consideration Table 5.11. The composition of the transaction costs for Canton Zurich leads to the following deductions:

- The overall costs of implementing the direct payment system amount to about CHF 4 million. Local authorities pay roughly one quarter of the transaction costs and the remaining costs are covered by the farmers.

- About one half of the transaction costs paid by local authorities are incurred at cantonal level, while the boroughs are liable for 38%. On the other hand, the State's share amounts to just about 10%.

- Almost half (45.6%) of the total transaction costs are attributable to the farms' records.

- The largest share of the transaction costs (roughly 24% each) is attributable to area payments and ecological compensation. In the case of both these measures, the farmers' records account for about 50% of the costs.

- Green Ticket measures generate a total of 60% of the transaction costs.

Two aspects must be taken into account when interpreting transaction costs at borough level: *a)* In the calculations, it is assumed that the boroughs expend three hours per farm (ALN estimate: 2 to 3 hours per farm). On the other hand, if an expenditure of two hours is assumed, costs decrease by about CHF 120 000 or 29%; *b)* a survey carried out in Canton Grisons indicated that the boroughs expend approximately three hours per farm. Given the higher share of arable land in Canton Zurich, it is most probable that the expenditure of boroughs there is, in general, greater than in Grisons.

The values of the key figures of transaction costs can be calculated on the basis of Table 5.11. The results for Canton Zurich are shown in Table 5.12.

- As a whole, the share of transaction costs in the direct payments disbursed amounts to 2.8%. In the basic variant, area payments followed by the other Red Ticket measures exhibit the highest efficiency with regard to the transfer of funds. In this respect, contributions for extensive cultivation and animal housing payments are the least efficient;

- Taking the whole area in Canton Zurich into consideration, transaction costs per relevant unit result in transaction costs of roughly CHF 56 per hectare. From this point of view, area-specific Green Ticket measures and slope payments are particularly beneficial. On the other hand, contributions for keeping animals under difficult conditions, ethological and area payments exhibit low transaction costs per unit;

- In the first instance, contributions for organic farming do less well if transaction costs are assigned to those farms taking part in a measure. This is due to the higher control costs paid by the farms. Generally speaking, average costs amount to about CHF 1 078 per farm and year.

Table 5.11. **Transaction costs in Canton Zurich (basic variant)**

CHF

Payer	State	Canton	Boroughs	Farm					Total	Share paid by authorities
Actor	State	Canton	Boroughs	Control organisations for organic farms	Control organisations for PEP farms	Records	Forms	Farm controls	Total	Share paid by authorities
Policy										
Area payments	25 804	151 225	108 916	3 572	99 248	445 255	56 879	31 814	927 713	285 946
Payments for keeping grazing farm animals	10 130	57 569	41 313	5 269	37 257	171 047	21 800	12 193	356 578	109 012
Payments for keeping livestock under difficult conditions	3 643	19 923	13 958	2 523	11 869	61 744	7 719	4 317	125 695	37 523
Payments for farming on steep slopes	4 990	27 636	19 658	2 632	16 164	77 538	10 653	5 958	165 230	52 284
Payments for ecological compensation	26 188	153 419	111 305	3 517	99 968	472 806	57 673	32 258	962 135	290 912
Payments for extensive cereal and rapeseed cultivation	10 574	58 691	42 379	3 756	37 633	173 669	21 857	12 225	360 785	111 644
Payments for organic crop farming	2 140	9 305	6 585	1 862	0	14 026	3 546	15 397	202 862	18 031
Payments for animal housing systems	6 295	34 059	24 382	1 700	64 090	134 631	12 800	13 604	301 562	64 737
Payments for turning animals outdoors regulary	10 640	60 540	43 138	2 020	107 120	248 161	22 836	27 118	539 574	114 318
Total Actor	100 404	572 368	411 635	21 852	473 349	1 798 877	215 763	154 885	3 942 134	1 084 407
Total Payer	100 404	572 368	411 635	2 857 726						

Table 5.12. **Key figures of transaction costs in Canton Zurich (basic variant)**

CHF

Policy	Total PRTC	Sum of payments	PRTC per payment (%)	Units	PRTC per unit	Sum of farms	PRTC per farm
Policy							
Area payments	927 713	93 458 643	0.9	ha	69 710	13.31	3 631	255.50
Payments for keeping grazing farm animals	356 578	13 259 231	2.8	LSU	15 546	22.94	1 819	196.03
Payments for keeping livestock under difficult conditions	125 695	3 994 000	3.5	ha	12 694	9.90	781	160.94
Payments for farming on steep slopes	165 230	2 423 925	6.8	ha	5 284	31.27	958	172.47
Payments for ecological compensation	962 135	12 749 724	7.5	ha	9 011	106.77	3 624	265.49
Payments for extensive cereal and rapeseed cultivation	360 785	2 546 774	14.7	ha	6 395	56.42	1 603	225.07
Payments for organic crop farming	202 862	2 092 912	9.0	ha	6 749	30.06	353	574.68
Payments for animal housing systems	301 562	2 616 491	11.5	LSU	24 855	12.13	1 166	258.63
Payments for turning animals outdoors regulary	539 574	7 586 914	7.1	LSU	42 660	12.65	1 956	275.86
Total	3 942 134	140 728 614	2.8	ha	70 990	55.53	3 657	1 077.97

Influence of the choice of method and assumptions on transaction costs (Canton Zurich)

The same procedure is used for classifying the basic variant for Canton Zurich as was chosen for Canton Grisons. Table 5.13 shows the key figures of the basic variant, the variants with modified labour costs as well as the variants for direct payment shares.

The costs of the labour costs variant are altered by CHF 1; the results show the influence of the alteration on the key figures. The results of Table 5.13 illustrate clearly that overall transaction costs rise by CHF 1.47 or 2.6% when labour expenditure increases by 5% or wages go up by CHF 1 per hour. On the whole, the choice of labour costs has the strongest influence on transaction costs for ecological compensation contributions (CHF 3.01 per unit) and payments for extensive cultivation (CHF 1.57 per unit). A rise in labour costs of CHF 1 per hour leads to an increase in transaction costs of roughly CHF 30 per farm.

Compared to the "labour costs" variant, the "direct payment share" variant has a much more pronounced influence on the key figures. However, it must be borne in mind that total transaction costs remain unchanged, as only the fixed items of implementation and control costs are allocated to the various measures according to a different key. When allocation takes place on the basis of direct payment shares, transaction costs related to area payments rise by about CHF 20 per hectare utilisable area, since they account for roughly two thirds of the total direct payments. On the other hand, transaction costs relating to other measures decrease, whereby there is a particularly marked decline in the costs of ecological compensation contributions and payments for extensive cultivation of cereals and oilseed.

To illustrate the results, the effects of both variants on the key figure PRTC per unit are shown in Figure 5.16. This figure also clearly demonstrates that changes in the key figures of the "direct payment share" variant are much more pronounced than in the case of the "labour costs" variant. The main reason for this is the fact that the "direct payment share" variant involves a shift away from the other measures towards area payments. As a result of this shift, transaction costs per franc of direct payments for all measures, with the exception of contributions for organic farming, are more or less evenly balanced. The value of the PRTC per unit key figure for area payments is similar to the value for organic farming contributions. Only the indicators for ecological compensation and payments for extensive cultivation are higher. The values for the other measures are significantly lower. The effect on the PRTC per farm indicator is even more striking. Area payments have the greatest value. Only the indicator for contributions for organic farming remains at a relatively high level; the values sink noticeably for the other measures.

To conclude the discussion of transaction costs in Canton Zurich, the shares of the individual measures in the overall transaction costs are illustrated in Figure 5.17. Once again, the darker lines indicate the basic variant, while the lighter line represents the "direct payment shares" variant. The influence of the choice of labour costs (CHF 0 to 25 per hour) is shown within the dark and light lines. Once again, the effects already discussed become apparent in the illustration:

- Only the share of the PRTC of the organic farming contribution measure is significantly influenced by the choice of labour costs. On the other hand, this share is hardly dependent on the choice of variant.

- The choice of the "direct payment share" variant results in a marked increase in the share of the PRTC related to area payments. The rest (with the exception of contributions for organic farming) decrease.

Table 5.13. **Influence of the choice of method and labour costs on the key figures of the transaction costs in Canton Zurich**

CHF

Policy	Units	Scenario								
		Base run			Influence of labour costs			Share of direct payments		
		PRTC per payment (%)	PRTC per unit	PRTC per farm	PRTC per payment (%)	PRTC per unit	PRTC per farm	PRTC per payment (%)	PRTC per unit	PRTC per farm
Policy										
Area payments	ha	0.99	13.31	25.50	0.03	0.37	7.09	1.60	21.47	412.20
Payments for keeping grazing farm animals	LSU	2.69	22.94	16.03	0.07	0.63	5.40	–0.56	–4.75	–40.56
Payments for keeping livestock under difficult conditions	ha	3.15	9.90	10.94	0.09	0.28	4.50	–0.92	–2.90	–47.16
Payments for farming on steep slopes	ha	6.82	31.27	12.47	0.19	0.85	4.71	–4.44	–20.38	–112.44
Payments for ecological compensation	ha	7.55	106.77	25.49	0.21	3.01	7.49	–4.76	–67.31	–167.37
Payments for extensive cereal and rapeseed cultivation	ha	14.17	56.42	25.07	0.39	1.57	6.26	–11.60	–46.20	–184.31
Payments for organic crop farming	ha	9.69	30.06	54.68	0.08	0.23	4.49	–0.39	–1.21	–23.23
Payments for animal housing systems	LSU	11.53	12.13	28.63	0.29	0.31	6.58	–6.35	–6.68	–142.42
Payments for turning animals outdoors regulary	LSU	7.11	12.65	25.86	0.19	0.34	7.33	–2.66	–4.74	–103.33
Total	ha	2.80	55.53	177.9	0.07	1.47	28.73	0.00	0.00	0.00

The influence of the scenarios "Influence of Labour Costs" and "Share of Direct Payments" is given as the difference to the scenario "Base run".

Figure 5.16. **Influence of the variants on the key figures PRTC per relevant unit (Canton Zurich)**

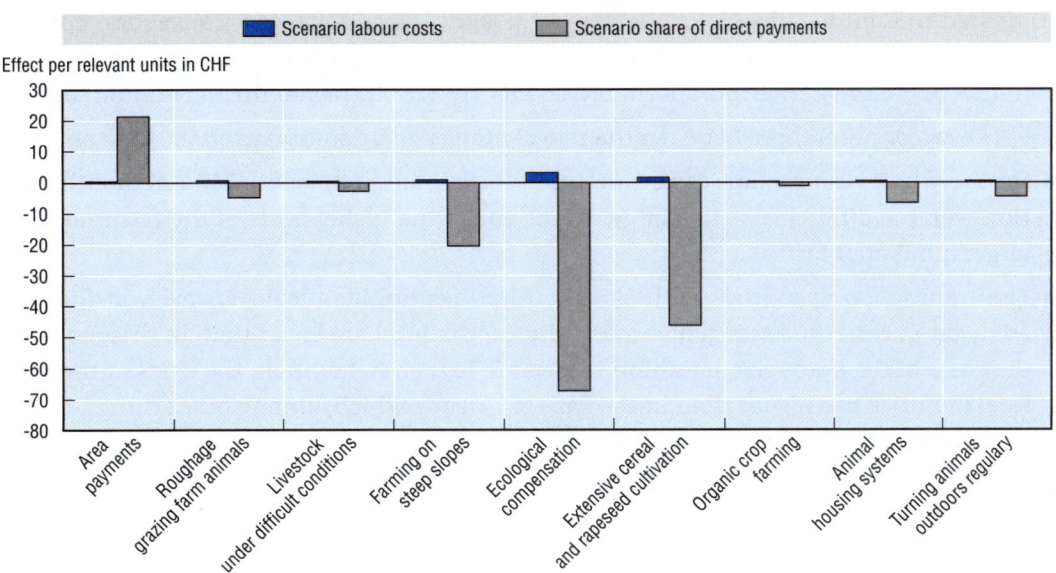

Figure 5.17. **Influence of the variants on the distribution of the PRTC according to measures (Canton Zurich)**

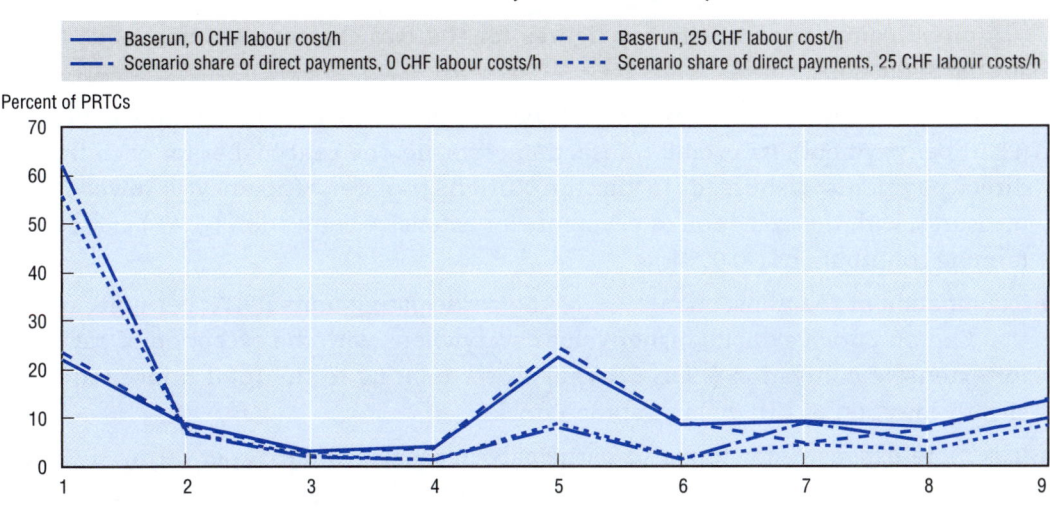

1. Area payments.
2. Payments for keeping grazing farm animals.
3. Payments for keeping livestock under difficult conditions.
4. Payments for farming on steep slopes.
5. Payments for ecological compensation.
6. Payments for extensive cereal and rapeseed cultivation.
7. Payments for organic crop farming.
8. Payments for animal housing systems.
9. Payments for turning animals outdoors regularly.

Differences between the case studies of Cantons Grisons and Zurich

The major differences relating to absolute transaction costs and key figures are discussed on the basis of the results obtained in the two case studies, whereby the differences as such are less important than their causes.

Table 5.14 illustrates the absolute cost differences between the Cantons Zurich and Grisons, whereby all values are to be understood as the difference of Canton Zurich as compared to Canton Grisons. In the case of positive deviations, the transaction costs in Canton Zurich are greater than in Canton Grisons. For example, the total costs for the boroughs in Canton Zurich are, absolutely, about CHF 128 000 higher than in Canton Grisons.

The major differences between the two cantons can be summarised in three points:

- On the whole, transaction costs in Canton Zurich are just about CHF 1 million higher. However, the difference is to a large extent due to the higher costs of the farms and the larger number of farms.

- The transaction costs covered by local authorities are identical. However, in the case study on Zurich, the costs generated at the canton level are lower than in Grisons while the costs of the boroughs in Canton Zurich are higher. In addition, the costs at the state level in Zurich are higher than in the case of Grisons which, among other things, can be explained by the larger number of farms.

- The deviations are relatively larger in the case of individual cost centres. This is due in part to methodological and system-specific reasons, as well as to structural and organisational differences between the Cantons. These are investigated in greater detail in the following.

Table 5.15 shows the key figures of the basic variant for the two case studies, as well as their differences.

A direct comparison of the key figures for the two case study areas leads to the following deductions:

- On the whole, Canton Zurich exhibits a lower degree of efficiency in the transfer of funds (PRTC per payment). Its overall transaction costs are one cent higher for each franc of direct payments disbursed. In Canton Zurich, transfer efficiency is lower for all measures, with the exception of payments for extensive cultivation (–2.4%) and organic farming contributions (–0.05%).

- Examination of the transaction costs of the participating units (PRTC per unit), reveals that Canton Zurich exhibits higher values everywhere, with the exception of payments for extensive cultivation (–9.85 CHF/ha). Costs relating to the total utilised area are slightly lower (–1.61 CHF/ha) in Canton Zurich.

- It is more expensive for farms to participate in measures (PRTC per farm) in Canton Zurich. This applies to all measures. However, generally speaking, the transaction costs generated by a farm which receives direct payments are, on average, about CHF 15 lower in Canton Zurich.

Analysis of factors influencing policy-related transaction costs

In this section the factors which influence transaction costs and the respective key figures are examined in greater detail. The influence factors are analysed on the basis of key figures for transaction costs obtained at farm level (basic variant; calculation according to Table 5.7). In order to avoid the methodological influences already discussed, the key figures of the transaction costs for the individual measures are no longer taken into account. Transaction costs are influenced by the following three factors:

- Factors relating to system and environment: In the first instance, system related influences involve the different direct payment rates which are based on agricultural zones. In particular, these influence the key figure "PRTC per direct payment". At the same time, the different rates and also the prevailing natural conditions influence the participation of farms in measures. The organic farming, BTS (animal housing systems) and RAUS (letting animals outdoors regularly) measures are excluded from the environment related factors as these are not restricted to a specific location. Furthermore, it is assumed that the conditions are homogeneous within the agricultural zones.

- Structural factors: Farm size is the most important structural factor which has a primary influence on the transaction costs a farm has to meet. Participation in organic farming, BTS and RAUS measures are regarded as further structural factors as they are not dependent on location. However, farm-specific structural indicators such as open arable land or areas devoted to special crops are not taken into consideration since, in addition to the orientation of the farm, these are primarily dependent on its location and are thus related to prevailing environmental conditions.

- Organisational factors: The organisational differences in implementation and controls between Cantons Grisons and Zurich are regarded as organisational factors. In addition to the factual organisational differences (differing "cost factor farms" for public authorities, different rates for control costs) the different fixed costs of the public authorities are also taken into account here.

Linear regression models are used to investigate the influence of these factors on the key figures of transaction costs at farm level. The purpose is not so much to obtain an exact quantification but rather to arrive at a qualitative determination of the influence factors. The influence factors are covered by the following variables.

System and environment related influences:

Zone i: Dummy variables for the zone location of a farm

Structural influences:

ALL_LN: Size of farm (in acres)

BIO: Dummy variable for organic farming

Raus: Dummy variable for participation in RAUS measure

BTS: Dummy variable for participation in BTS measure

Organisational influences:

Canton: Dummy variable for cantonal location (1 = Canton Zurich)

This results in the following general linear regression model:

$$\text{Transaction costs per farm} = C + a * \text{Canton} + b_i * Z_i + c * \text{BIO} + d * \text{Raus} + e * \text{BTS} + f * \text{ALL\_LN} + u$$

The model parameters can be interpreted as follows:

- Constant C: average transaction costs for proof of ecological performance; the costs of the Green Ticket measures are covered by the dummy variables.

- Parameter a: cost difference between the Cantons due to different organisation of controls and implementation.

Table 5.14. **Differences between the cantons with regard to absolute transaction costs (Deviation of Canton Zurich compared to Canton Grisons)**

CHF

Payer	State	Canton	Boroughs	Farm					Total	Share paid by authorities
Actor	State	Canton	Boroughs	Control organisations for organic farms	Control organisations for PEP farms	Records	Forms	Farm controls	Total	Share paid by authorities
Policy										
Area payments	16 649	46 338	63 513	-16 607	70 152	306 578	30 820	17 238	534 681	126 500
Payments for keeping grazing farm animals	1 533	-38 324	-269	-19 442	12 370	43 756	-1 921	-1 074	-3 372	-37 061
Payments for keeping livestock under difficult conditions	-5 023	-75 972	-27 624	-22 233	-12 977	-67 095	-16 003	-8 951	-235 876	-108 618
Payments for farming on steep slopes	-3 556	-66 941	-21 293	-21 924	-8 289	-46 907	-12 770	-7 142	-188 822	-91 790
Payments for ecological compensation	17 176	39 630	66 696	-16 443	71 652	333 172	32 105	17 957	561 945	123 501
Payments for extensive cereal and rapeseed cultivation	9 663	38 732	38 251	1 795	34 920	154 015	19 580	10 951	307 907	86 646
Payments for organic crop farming	-3 811	-62 305	-12 797	-204 042	0	-37 041	-7 586	-43 573	-371 155	-78 913
Payments for animal housing systems	3 984	-6 229	14 188	3 053	60 497	101 570	7 072	6 665	190 801	11 943
Payments for turning animals outdoors regulary	3 230	-41 929	7 348	-4 830	88 970	94 465	2 511	-4 026	145 739	-31 351
Total Actor	39 844	-167 001	128 013	-300 672	317 294	882 514	53 808	-11 954	941 847	857
Total Payer	39 844	-167 001	128 013			940 990				

All values are to be understood as the difference of Canton Zurich compared to Canton Grisons. Values greater than zero represent higher transaction costs in Canton Zurich than in Canton Grisons.

Table 5.15. **Differences between the cantons with regard to the key figures**

CHF

Policy	Units	Zurich			Grisons			Difference		
		PRTC per payment (%)	PRTC per unit	PRTC per farm	PRTC per payment (%)	PRTC per unit	PRTC per farm	PRTC per payment (%)	PRTC per unit	PRTC per farm
Area payments	ha	0.99	13.31	255.50	0.63	7.52	143.44	0.37	5.79	112.06
Payments for keeping grazing farm animals	LSU	2.69	22.94	196.03	1.21	9.88	136.81	1.48	13.06	59.22
Payments for keeping livestock under difficult conditions	LSU	3.15	9.90	160.94	0.94	8.98	137.22	2.21	0.92	23.72
Payments for farming on steep slopes	ha	6.82	31.27	172.47	2.52	11.04	136.86	4.30	20.23	35.62
Payments for ecological compensation	ha	7.55	106.77	265.49	6.75	27.49	147.62	0.79	79.28	117.87
Payments for extensive cereal and rapeseed cultivation	ha	14.17	56.42	225.07	16.57	66.26	195.12	-2.41	-9.85	29.95
Payments for organic farming	ha	9.69	30.06	574.68	9.74	20.06	413.56	-0.05	10.00	161.12
Payments for animal housing systems	LSU	11.53	12.13	258.63	8.32	8.05	148.47	3.20	4.09	110.16
Payments for turning animals outdoors regulary	LSU	7.11	12.65	275.86	5.02	9.01	167.59	2.09	3.64	108.27
Total	ha	2.80	55.53	1 078.00	1.80	57.14	1 093.00	1.00	-1.61	-15.03

The values in the column "Difference" are to be understood as the difference of the key figures of Canton Zurich compared to Canton Grisons. Values greater than zero represent higher indicators for Canton Zurich than for Canton Grisons.

- Parameter b_i: cost differences according to zones.
- Parameters c, d, e: influence of participation in organic farming and ethological programmes (Green Ticket measures).
- Error term u: other influences, *e.g.* farm-specific influences.

PRTC *per farm*

The results of the regression for the total transaction costs per farm are shown in Table 5.16, whereby the regression involves all the farms in the two Cantons. By and large, the regression explains just about 80% of the variance in transaction costs per farm ($R^2 = 0.797$). The significance level of the coefficients reveals that all the variables used in the regression have a highly significant influence on transaction costs. On average, a farm which is eligible to receive direct payments in Canton Grisons in mountain zone 3 (Zone 53) generates transaction costs amounting to CHF 538. These costs represent average implementation and control costs for proof of ecological performance.

Table 5.16. **Influence factors on transaction costs per farm**

Dependant variable: Total PRTCs per farm

Model	Unstandardised coefficients		Standardised coefficients		
	B	Std Error	Beta	t	Sig.
Constant	538.0	6.3		85.8	0.00
Canton	−97.9	8.7	−0.143	−11.2	0.00
BIO	−49.6	5.1	−0.065	−9.8	0.00
RAUS	100.0	1.0	0.100	23.1	0.00
BTS	55.7	4.7	0.075	11.9	0.00
Zone 11	82.1	10.7	0.098	7.7	0.00
Zone 21	95.7	9.9	0.088	9.6	0.00
Zone 22	106.7	10.8	0.117	9.9	0.00
Zone 41	117.0	11.0	0.092	10.6	0.00
Zone 51	42.1	12.0	0.025	3.5	0.00
Zone 52	50.1	10.8	0.029	4.7	0.00
Zone 54	−34.9	6.3	−0.037	−5.5	0.00
All_LN	0.2	0.0	0.822	140.8	0.00

Organisational influences (cantonal location) amount to roughly CH 98. On average, the transaction costs generated by a farm in Canton Zurich are just about CHF 100 lower compared to a farm in Canton Grisons.

Costs vary compared to the basic value of CHF 538 (farm in Grisons in Zone 53) by CHF −35 to CHF 117 depending on the zone in which a farm is located. Farms in zones which are less favourable for agricultural production (especially mountain zones) exhibit lower costs.

The results of the regression indicate that participation in the Green Ticket measures RAUS and BTS raises costs while, on average, the costs generated by organically run farms are lower. The fact that organic farms exhibit lower costs than PEP farms is due to lower requirements regarding the keeping of records. This reduces the farm-specific expenditure of organic farms and, on average, offsets the higher control costs.

On average, costs increase by CHF 0.24 per or by CHF 24 per hectare in relation to the utilised agricultural area of a farm.

If the payers of PRTC per farm, namely the farm and public authorities, are considered separately the following can be observed:

- Compared to the costs incurred by the farms, those arising for public authorities are virtually independent of zones and are thus not affected by factors related to the system or environment.

- On the other hand, the costs incurred by the farms are, to a large extent, independent of cantonal location.

- The factual fixed costs of the farms amount to about CHF 164 per farm (for Zone 53). The total fluctuation range associated with this value lies between CHF –31 (Zone 54) and CHF 105 (Zone 41). The remainder of constant C (roughly CHF 373) in Table 5.16 represents the fixed costs plus the costs arising for the public authorities from the number of farms. This is spread over the farms.

- Farms taking part in the organic farming measure generate slightly higher costs for public authorities. On the other hand, these farms have lower transaction costs.

PRTC per unit

The PRTC per unit key figure is generated by dividing the total transaction costs of a farm by its area. This means that the key figure depends notably on the size of the farm since the fixed cost share is spread over a greater area in the case of larger farms. On the other hand, variable costs (about CHF 24/ha) are attributed to the indicator as a fixed sum (due to the division by the area).

The effect of farm size on transaction costs per area unit is illustrated in Figure 5.18 for farms with an area of between 10 and 60 ha. The estimated inverse function (PRTC per $a = c_0 + c_1 /$ ALL_LN $+ u$) explains about three quarters of the variance of the transaction costs per area unit (R^2: 0.732).

Generally speaking, the size of the farm does not exhibit any special dependence on the influence factors. Therefore basically, the statements relating to the PRTC per farm key figure can be applied to the interpretation of the PRTC per unit key figure (exception: area dependence). However, the dependencies discussed in relation to the PRTC per farm key figure are lower for the PRTC per unit than the scales of size shown in the Figure.

Figure 5.18. **PRTC per unit of area depending on farm size**

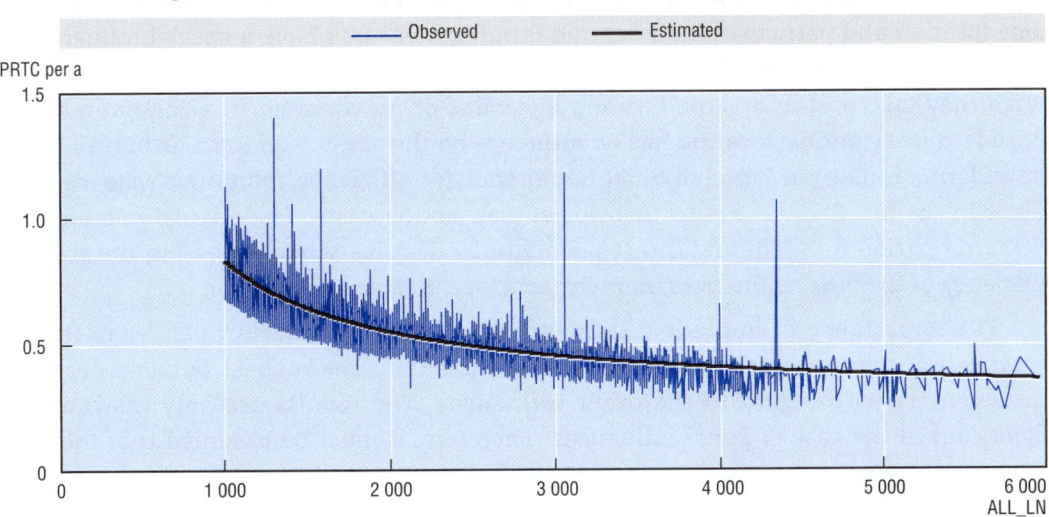

PRTC per direct payment

The PRTC per direct payment key figure is obtained by dividing a farm's absolute transaction costs by its total direct payments. On the one hand, the amount of the direct payments depends markedly on the direct payment system (different rates for different zones), and on the other hand is also subject to environmental conditions (e.g. intensities for grazing LSU payments) and farm management and orientation (e.g. share of ecological compensatory areas). Farm size also has a great influence as all direct payments are linked either directly or indirectly to the land area involved. In addition, the rates of certain payments are reduced depending on the size of the farm (e.g. reduction of area payments for farms with over 30 ha; see Figure 5.2). Determination of the amount of a farm's direct payments (Table 5.17, $R^2 = 0.831$) confirms the importance of system and environment related influences. Farms at higher altitudes (Zones 51 to 54) receive noticeably larger direct payments than farms in the lowlands (e.g. Zone 11) due to specific measures designed to benefit mountain areas. Since the additional services of Green Ticket measures are subject to special payments, participation in these programmes leads to a significant increase in direct payments.

Table 5.17. **Dependency of direct payments per farm**
Dependant variable: Direct payments per farm

Model	Unstandardised coefficients		Standardised coefficients		
	B	Std error	Beta	t	Sig.
Constant	18 067.4	496.8		36.4	0.00
BIO	5 396.8	401.6	0.084	13.4	0.00
RAUS	4 061.4	382.0	0.066	10.6	0.00
BTS	7 954.1	370.2	0.127	21.5	0.00
All_LN	18.4	0.1	0.739	135.3	
Zone 11	−21 556.3	509.9	−0.304	−42.3	0.00
Zone 21	−20 338.8	584.4	−0.220	−34.8	0.00
Zone 22	−21 721.1	517.0	−0.282	−42.0	0.00
Zone 41	−17 716.3	645.0	−0.163	−27.5	0.00
Zone 51	−13 035.2	793.1	−0.091	−16.4	0.00
Zone 52	−7 691.5	814.9	−0.052	−9.4	0.00

An understanding of the influence values on the PRTC per farm reveals clearly that zone location and participation in organic farming measures have a special influence on transfer efficiency. Farms which are otherwise identical exhibit higher transfer efficiency when they take part in organic farming measures or are situated in a zone at a higher altitude. Since cantonal location has no influence on the direct payments disbursed by the State, farms in Canton Zurich exhibit better transfer efficiency than otherwise identical farms in Canton Grisons due to their lower transaction costs. The change in additional PRTC in relation to additional direct payments is decisive when assessing the transfer efficiency of the other influences, namely farm size, BTS and RAUS.

The importance of individual influences for transfer efficiency can be examined separately if the farms selected for investigation exhibit the highest possible degree of homogeneity with regard to the other influences. The results are only relevant and significant in the case of zonal influences. Therefore, it must be assumed that the farm management and orientation (*e.g.* different branches or ecological area shares) and farm-specific conditions also have a marked influence on transfer efficiency.

On the whole, the sum of the transaction costs and their controlling factors have very little influence on transfer efficiency. It is rather farm management and orientation as well as the direct payment system which are decisive for the determination of the sum of the total payments granted to a farm.

5.5. Conclusions

The transaction costs of the Swiss direct payment system are estimated using the methods presented in this chapter. In this way, implementation and control costs are obtained for the various levels, namely the state, cantons, control organisations, boroughs and farms. While the costs arising for public authorities and control organisations can be determined with exactitude, there are uncertainties on the farms themselves with regard to labour expenditure for keeping records and completing forms as well as for labour costs. The influence of these uncertainties on total transaction costs is assessed by means of sensitivity analyses in order to test the validity of the estimated costs.

Transaction costs amounting to roughly CHF 3.0 million and CHF 3.9 million are estimated for, Grisons and Zurich, respectively. Table 5.18 shows the average transaction costs per farm, per hectare utilised area and per franc of direct payments granted for the two cantons. In both cantons, implementation and control costs amount to roughly CHF 1 100 per farm. In Canton Grisons, public authorities pay just about 37% and in Canton Zurich 30% of this sum. Transfer efficiency varies between 1.8% and 2.8%, whereby the superior efficiency in Canton Grisons is due primarily to higher direct payments.

Table 5.18. **Estimation of transaction costs for the case study cantons**

	Total PRTC	PRTC paid by authorities	PRTC per farm	PRTC per hectare	PRTC per direct payment
Grisons	3.0 Mio. CHF	1.1 Mio. CHF	1 094 CHF	55.5 CHF	1.8%
Zurich	3.9 Mio. CHF	1.1 Mio. CHF	1 078 CHF	57.1 CHF	2.8%

Absolute transaction costs and the key figures are influenced by various factors. In the case of the key figures, this applies in particular to the allocation of transaction costs to individual measures and the associated methodological uncertainties. This is due to the fact that only a small part of the processes which are relevant to the costs can be allocated directly to the measures. Therefore, statements concerning transaction costs and their influence factors which are based on cost comparisons between the individual measures must be viewed with certain reservations. However, total transaction costs do not depend on this allocation. The influence factors which are more decisive for the amount of the transaction costs are:

- farm size;
- farm's participation in green ticket measures;
- organisational differences between the cantons and consequently the location of the farm within one canton or the other;
- orientation of the farm;
- environmental influences.

The sum of the transaction costs per participating unit depends primarily on the size of the farm. Bigger farms can spread the fixed cost share of their transaction costs over a larger area and therefore generate lower transaction costs per hectare.

On the other hand, transfer efficiency (PRTC per direct payment) depends largely on influences related to the system and the environment, plus the farm's orientation and the associated direct payments. In this case, the factual transaction costs only play a subordinate role and the direct payments disbursed per farm are far more significant.

Eligibility to receive direct payments is linked to precise regulations at farm level. This means that transaction costs are transparent for the farmers and, in addition to the individual capabilities of the farm manager, can be attributed directly to the processes stipulated by the State. Consequently, a farm's costs can only be reduced if the regulations and processes are optimised in relation to the desired quality and the costs of the farm.

There must be a direct relationship between the allocation and interpretation of transaction costs and the respective direct payments programmes and agricultural policy target system. Two different aspects must be considered:

● If the direct payment system is viewed as a system for die provision of those services defined for agriculture under the terms of the Federal Constitution, then transaction costs can be interpreted as part of the costs of quality assurance within this system. The largest part of these costs is attributable to controls of the regulations governing eligibility to receive payments. In the case of general direct payments, these regulations cover fulfilment of proof of ecological performance (cross compliance) as well as special rules for payments for organic farming and ethological contributions. The direct payments granted also represent part of costs of remunerating agriculture for the provision of specific services. This remuneration amounts to about CHF 307 million over both cantons. The implementation and control of these services cost public authorities roughly an additional CHF 2 million, or 0.7% of the direct payments disbursed. Therefore, from the point of view of the public authorities, transaction costs can be regarded as very efficient. It is possible that there is a slight potential for a reduction of the costs incurred by public authorities, as can be seen from the differences between the cantons. Approaches to realise this potential without loss of quality involve far-reaching simplification of the processes and quality requirements.

● On the other hand, if the direct payment system is interpreted as a system purely designed for the transfer of income, then transfer efficiency can be improved by simplifying the system and, in particular, the regulations. However, this argument neglects to consider the public services which agriculture is bound to supply under the terms of the Federal Constitution and which will hardly be provided by means of a mere transfer of income. This applies in particular to quality-specific targets, the promotion of positive services and the avoidance of negative externalities in agricultural production. According to current agricultural legislation, these public goods are remunerated by means of direct payments, whereby a part of these payments serve to support agricultural incomes.

Regardless of the form of agricultural support (e.g. remuneration for services or price support), it can be assumed that the amount of the transaction costs is primarily attributable to the desired quality of the public goods, i.e. the multifunctional services provided by agriculture. This applies to both public authorities and the farms themselves. Given today's direct payment system, transaction costs can only be reduced significantly by adapting the quality requirements relating to the multifunctional services. Improvements in implementation and control efficiency demand simultaneous optimisation of transaction costs and the quality of the services, whereby these two dimensions have conflicting objectives.

References

BLW, Bundesamt für Landwirtschaft (1998), Direktzahlungen an die Landwirtschaft, Bericht der Hauptabteilung Direktzahlungen und Strukturen des BLW, Bern.

BLW, Bundesamt für Landwirtschaft (2003a), AGIS-Datenbank, Bern.

BLW, Bundesamt für Landwirtschaft (2003b.), Agrarbericht 2003 des Bundesamtes für Landwirtschaft (Agricultural Report 2003), Bern.

BLW, Bundesamt für Landwirtschaft (2004), Ökologischer Leistungsnachweis (Proof of ecological performance), Bern (www.blw.admin.ch/rubriken/00453/index.html?lang=de, Status as per 20 December 2004).

BLW, Bundesamt für Landwirtschaft (div. Jg.), Agrarberichte des Bundesamtes für Landwirtschaft (Agricultural Reports), Bern.

BLW, Bundesamt für Landwirtschaft (div. Jg.), Direktzahlungen an die Landwirtschaft (Report on the disbursement of direct payments), Berichte der Hauptabteilung Direktzahlungen und Strukturen des BLW, Bern.

Christensen, T. and H. Rygnestad (2000), Environmental Cross Compliance: Topics for Future Research, Frederiksberg (www.foi.dk/Publikationer/wp/2000-wp/wp200001.pdf, 22 December 2004).

European Environment Agency (2004), Definition of Cross Compliance, Copenhagen (http://glossary.eea.eu.int/EEAGlossary/C/cross-compliance, Status as per 15 November 2004).

Huber, P. (1998), Die Verwaltungskosten des Agrarsystems, Wien: AK.

Mann, S. (2001), "Zur Effizienz der deutschen Agrarverwaltung", Agrarwirtschaft, Vol. 50, issue 5, pp. 302-307.

OECD (2004), Agricultural Policies in OECD countries: At a Glance, OECD, Paris.

Rieder, P. (1998), Auswirkungen eines EU-Beitritts auf die schweizerische Agrarpolitik und Landwirtschaft, Schriftenreihe Institut für Agrarwirtschaft ETH Zürich 2/1998, Zürich.

Rieder, P., C. Flury and G. Giuliani (2003), Estimation du soutien à l'agriculture: Alternative à la présentation traditionnelle de l'ESP, Institut d'Économie Rural, groupe marché et politique, École polytechnique fédérale de Zürich, Zurich.

ISBN 978-92-64-03091-6
The Implementation Costs of Agricultural Policies
© OECD 2007

PART II

Chapter 6

A Case Study of Policy-related Transaction Costs in Land Conservation Programmes in the United States

Executive Summary

The Conservation Reserve Program (CRP) has been the largest land retirement programme, and the largest US conservation programme of any kind. CRP is a voluntary programme that offers annual rental payments, incentive payments for certain activities, and cost-share assistance to establish approved cover on eligible cropland.

As of June 2004, there were 34.8 million acres enrolled in CRP, with 2.8 million of those acres in partial-field enrolments under the continuous signup, Conservation Reserve Enhancement Program (CREP), and the Farmable Wetland Program (FWP). More than 660 000 contracts were in force with more than 390 000 farmers covering this acreage. The annual rental cost was almost USD 1.7 billion, and the average rental cost per acre was USD 48. About 60% of CRP acreage was planted to grasses, 16% to trees or woody vegetation for wildlife, and 5% was dedicated to wetland restoration. The programme is authorized at 39.7 million acres for enrolments under 10-15 year contracts through 2007. The current CRP program targets land retirement to increase the cost-effectiveness of the programme, and adjusts the payment to closely match the market value.

Responsibility for implementing CRP is shared by two of the US Department of Agriculture's largest agencies. Educating and providing technical assistance to farmers rests with USDA's Natural Resources Conservation Service (NRCS), while financial assistance paid to farmers is administered through USDA's Farm Service Agency (FSA). Other USDA, Federal, State and local government agencies have also played roles in developing and implementing conservation programmes, but their participation is usually short-lived *ad hoc* a minor financial commitment, and not easily captured in agency budget accounts.

The primary form of technical assistance provided in CRP is preparation of the conservation plan for CRP acres. An approved conservation plan is required before a CRP contract can be approved. Before approving CRP contracts, the FSA County Committee reviews and approves the plan to ensure that the plan meets all requirements. FSA has responsibility for issuing and implementing the contract for financial assistance, and making payment. The approved conservation plan obligates CRP participants to establish and maintain approved practices. NRCS has responsibility for monitoring and enforcing technical aspects of the plan's cover and practice establishment and maintenance, while FSA ensures that other contract terms continue in force.

Overall, the costs to the government of implementing the CRP are relatively low, running from 3% of expenditures in initial years and 1% in succeeding years for NRCS technical assistance, and about 4% of expenditures for FSA administrative support costs. This amounts to about USD 60 per acre enrolled in initial years of a ten-year enrolment period, and about USD 20 per acre in succeeding years. These costs are less than comparable costs for the Wetland Reserve Program, and much less for working land programmes such as Environmental Quality Improvement Program (EQIP) and its predecessor programmes. The absolute size of rental payments in CRP dwarfs transaction costs in ways that cost-share funds under working lands programmes do not.

FSA administrative costs are highly correlated with programme characteristics, especially the cumulative acreage enrolled, with each additional acre increasing costs by USD 1.79. NRCS technical assistance costs are more variable, and are significantly correlated with acres idled or installed in a given year (adding USD 2.39 per acre) and cumulative acres enrolled (adding USD .30 with each additional acre) in each year. NRCS costs increased significantly between the first CRP signups and the second set after 1996. The signs and magnitudes of other correlates are interesting, but not statistically significant.

Conservation technical assistance overall has declined from peak levels in the mid-1970s, despite an increase in management-related practices. Congressional support for technical assistance may be dropping, as evidenced by issues related to the Section 11 cap on reimbursement, caps on technical assistance in the new Conservation Security Program (CSP), reliance on third-party technical assistance providers, and mandates for studying conservation planning reform.

Information technology, centralisation of functions, and other administrative improvements can reduce technical assistance and administrative transaction costs, and perhaps improve the ability to evaluate resource concerns and conservation plans to correct them, and has partly compensated for reduced technical assistance availability in recent years. In particular, FSA has invested in distributed, web-based signup processes and software that have substantially reduced the burden of signup on their local county offices. Continued decreases in technical assistance at the field level and reliance on online resources and information technology can and have made some cost reductions, but cannot indefinitely substitute for face-to-face, on-the-ground technical assistance provided by trained conservationists to producers. Technical assistance is not merely a policy-related transaction cost to be overcome, but part of the programme itself.

Acronyms for USDA conservation agencies and programmes

Acronym	Description
ACP	Agricultural Conservation Program (FSA)
ASCS	Agricultural Conservation and Stabilisation Service, now FSA
CCC	Commodity Credit Corporation
CD	Conservation District
CED	County Executive Director (FSA)
C.F.R.	Code of Federal Regulations
COC	County Committee (FSA)
CPA	Conservation Priority Area in CRP
CRBSCP	Colorado River Basin Salinity Control Program
CRP	Conservation Reserve Program (FSA)
CREP	Conservation Reserve Enhancement Program (FSA)
CSP	Conservation Security Program (CSP)
CTA	Conservation Technical Assistance Program (NRCS)
CWA	Clean Water Act
CZARA	Coastal Zone Act Reauthorisation Amendments
DC	District Conservationist or Designated Conservationist
EBI	Environmental Benefits Index in CRP
ECP	Emergency Conservation Program (FSA)
EI	Erodibility Index

Acronyms for USDA conservation agencies and programmes (cont.)

Acronym	Description
EQIP	Environmental Quality Incentives Program (FSA)
EPA	Environmental Protection Agency
FACTA	Food, Agriculture, Conservation, and Trade Act of 1990
FAIR	Federal Agriculture Improvement and Reform Act of 1996
FIFRA	Federal Insecticide, Fungicide, and Rodenticide Act
FIP	Forestry Incentives Program (FS)
FRPP	Farm and Ranchland Protection Program (NRCS)
FSA	Farm Service Agency
1985 FSA	Food Security Act of 1985
FS	US Forest Service in USDA
FSRI	Farm Security and Rural Investment Act of 2002
FWP	Farmable Wetlands Program (FSA)
FWS	US Fish and Wildlife Service in USDOI
GAO	General Accounting Office
GIS	Geographic information system
GPCP	Great Plains Conservation Program
GRP	Grassland Reserve Program (FSA)
HEL	Highly Erodible Land
MARR	Maximum Acceptable Rental Rate in CRP before 1990
NACD	National Association of Conservation Districts
NEPA	National Environmental Policy Act of 1970
NPPH	National Planning Procedures Handbook
NRCS	Natural Resources Conservation Service
OBPA	Office of Budget and Program Analysis
OECD	Organisation for Economic Cooperation and Development
PRTC	Policy-related transaction costs
RCWP	Rural Clean Water Program (NRCS)
SCS	Soil Conservation Service, now NRCS
SRR	Soil adjusted rental rate in CRP after 1990
T&E	Threatened and endangered species
TSP	Technical Service Provider provision
USDA	US Department of Agriculture
USDOI	US Department of the Interior
WBP	Water Bank Program
WHIP	Wildlife Habitat Incentives Program
WQIP	Water Quality Incentives Program
WRP	Wetland Reserve Program

6.1. Background

This case study examines the evolution of land conservation programmes in the United States and provides estimates of technical assistance expenditures and other implementation costs for the Conservation Reserve Program (CRP), the Wetland Reserve Program (WRP) and the Environmental Quality Improvement Program (EQIP) and their predecessors.

Policy-related transaction costs and technical assistance

This chapter is a case study of the US Conservation Reserve Program (CRP, see Appendix for Acronyms), in the context of programme delivery for other US Department of Agriculture (USDA) conservation programmes. The primary focus of the case study is on the levels of and trends in administrative and technical assistance costs incurred by USDA agencies implementing CRP and other conservation programmes. However, before launching into that examination, it is useful to consider how well these costs really fit the economist's concept of transaction costs, definition of PRTC's adopted in Chapter 1, and the extent to which they can even be logically separated from delivery of the programmes they are intended to implement.

Economists use the term "transaction costs" to mean many things, ranging from very specific kinds of charges associated with narrow market transaction in particular markets to very broad categories of expense associated with corporate structure and industrial organisation.[1] Chapter 1 is using the term in a narrower sense to encompass only administrative or implementation costs associated with a particular programme. Both of these senses, however, have a pejorative connotation that implies transaction costs are a kind of "friction" on the system that should be eliminated or minimised. The underlying question addressed in this topic shows that transaction costs are invariably considered as negative, since they are seen as offsetting economic benefits.

This attitude toward transaction costs, combined with an understandable reaction to bureaucratic excesses in some kinds of programmes, leads policy makers to prefer market-based policies whose efficiency seems to be determined by proportion of expenditures that can be directly paid to producers. In the case of agri-environmental programmes, not all costs of implementing the programme decrease cost-effectiveness, and some may be as valuable (or more valuable) than the financial assistance paid for the "practice". This should be clear if we enumerate the categories of PRTCs identified in Figure 1.1 of Chapter 1, which occur in the CRP program.

Initial and final costs

While a certain amount of agency staff time was expended in designing the modern CRP in 1985 (and redesigning it after the 1990, 1996 and 2002 Farm Bills), much of the effort was undertaken by a small number of policy specialists whose job is designing and providing input to such programmes, and is essentially a fixed (and sunk) cost, although the opportunity costs of research foregone by those specialists is a real, but small, cost. Many of the participants in the design process actively engaged in the trade-off between requiring more elaborate environmental and economic bid assessment and reducing administrative burdens on the agencies they represented in implementing it. Likewise, such *ex-post* evaluations of programme performance as have occurred are grist for the political process of comparing expenditures on one programme to another, on agri-environmental programmes to other environmental policies, and on comparing agricultural programmes to other functions of government. These activities contribute to the overall efficiency of the governance process and are not fairly counted to just the programme itself. Only those expenditures within the primary agencies responsible for the programme are accounted for in this case study.

Implementation costs

Most of the administrative data on programme expenditures fits into this category, and it is reasonably comprehensive for the primary agencies involved in running CRP. Some of these steps (filling out forms, cutting checks, etc.) can and should be accomplished as efficiently as possible and share in benefits from centralisation and decentralisation in USDA-wide reorganisations accomplished in 1994 that consolidated such administrative functions across the entire department (GAO, 2000). For example, USDA's National Finance Center processes checks for CRP participants as only part of the complex of farm programme payments and financial transaction (payroll, personnel, etc.) for all USDA and many other Federal agencies. USDA's traditional decentralisation of field offices to the county level accomplishes administrative efficiencies because Farm Service Agency (FSA) and Natural Resources Conservation Service (NRCS) staffing for CRP is only part of a full slate of administrative and technical services provided for conservation and other farm programmes. Having county offices also reduces producers' transaction costs (travel, communication, information) that would be imposed by a more centralised staffing.

Technical assistance costs are a larger cost of implementing CRP that combines aspects of transaction costs with the substance of the programme itself. Identifying and targeting resources to be protected and developing the conservation cover plans needed to accomplish that protection are tasks that require trained resource professionals working in close concert with the producer on the land. While modern information technology can reap efficiency rewards in some cases, excessive reliance on information technology could result in poorly targeted and ill-planned land retirement.

Previous USDA land retirement programmes (the original CRP and Soil Bank) in the 1950s and 1960s were not targeted to specific agri-environmental resources (highly erodible land, wetlands, water quality, wildlife habitat), nor was conservation cover planning aimed at correcting or improving specific resource issues. While it is not possible to quantitatively compare the environmental impacts of current and earlier programmes because earlier programme impacts were not studied and because of systemic changes in agricultural production techniques and markets, it is generally believed that environmental benefits from current land retirement are far greater than in the past.

Participation costs

There is little data on the costs producers incurred in enrolling land in CRP, with the exception of some small-scale surveys of producer satisfaction with the signup process. While it is certainly desirable to minimise the administrative burden of applying for and enrolling in the programme, it can be argued that the time producers spend working with resource specialists on technical assistance in selecting parcels to enrol and developing conservation cover plans has value above its necessity for implementing the programme. Indeed, another long-standing NRCS conservation program, the Conservation Technical Assistance (CTA) program provides nothing but staff time to interact with producers, and much of the State-Federal Cooperative Extension effort in conservation is similarly devoted to working one-on-one with producers. While efficiencies should be sought in delivering this part of the programme, the goal should be to enrich the time spent in targeting and planning, not eliminating it.

Plan of the chapter

The balance of this chapter covers the following areas:

- A brief description of the CRP program.
- The evolution from whole-farm conservation planning to programme-based technical assistance.
- The evolution of the interagency roles of FSA and NRCS in administering CRP and other conservation programmes.
- The differentiation between costs of developing and administering programmes such as CRP, technical assistance to producers for applying for the programme, and technical assistance in applying conservation on the ground once an application is approved.
- The interplay between open enrolment, complex evaluation of applications, and cost-effectiveness in administering CRP.
- Comparing and contrasting technical assistance expenditures for annual rental programmes like CRP, long-term easement programmes like WRP, and conservation on working lands like the former ACP and current EQIP programmes.
- Trends in the willingness to fund technical assistance (*versus* rental payments, cost-share, etc.) when the trend in conservation is toward more and more "management" practices that require even more technical assistance.

The analysis is conducted using budget data on administrative support, technical assistance, cost-sharing, and rental payments for CRP and other USDA conservation programmes from 1983 to 2002 (USDA, OBPA, 2002). A longer time series on conservation expenditures from 1936 to 1999 is also used to illustrate longer-term trends in technical *versus* financial assistance.

6.2. The Conservation Reserve Program

Land retirement has been a common agricultural policy tool in the United States since the 1930s, when dual concerns over low farm income and resource problems, such as soil erosion, flooding, and drought, were both addressed by reducing cultivated acreage. The United States has periodically instituted programmes to idle cropped acreage, with peak enrolments varying inversely with net farm income. Historically, these programmes have been instituted when agricultural prices were low, and acreage came out of enrolment when prices recovered. Land retirement programmes in the US (including annual set-asides) averaged 31 million acres between 1933 and 2001, (8% of cropland used for crops; Figure 6.1). Land retirement ranged as high as 78 million acres (20%) in 1983. In only 10 years (1948-55 and 1980-81) was no cropland retired in such programmes.

CRP has been the largest land retirement programme throughout this period, and the largest US conservation programme of any kind. CRP is a voluntary programme that offers annual rental payments, incentive payments for certain activities, and cost-share assistance to establish approved cover on eligible cropland. One potential difference between the United States and many other countries is that in the United States, private landowners have almost no limitations on how they can use their land for agricultural or

Figure 6.1. **History of US land retirement programmes, 1933-2001**

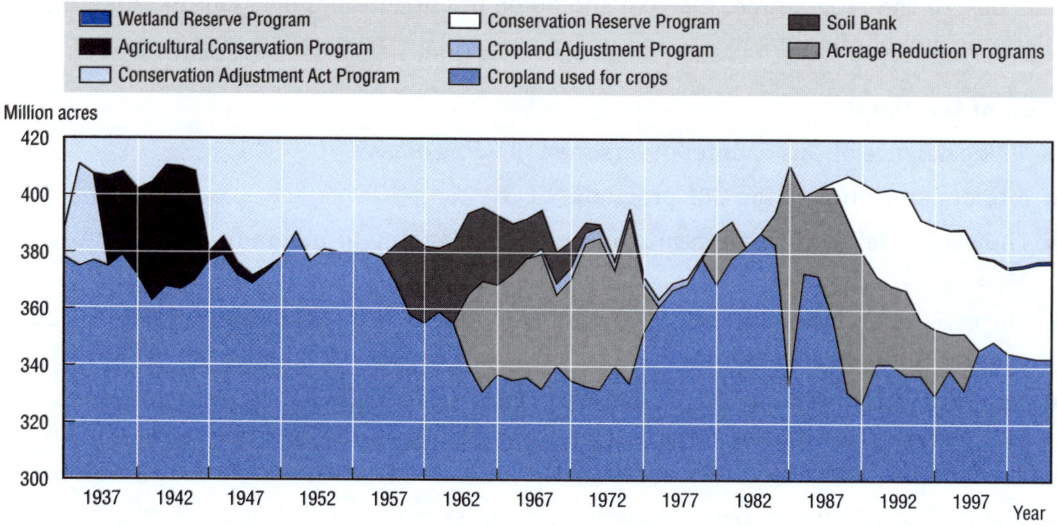

Source: Heimlich (2002).

forestry purposes. There are no national laws, and only a few State or local regulations that significantly restrict landowners' use of their land (Taylor, 2001). Consequently, any attempt to influence land use or adoption of conservation practices has to rely on direct voluntary incentives, or indirect incentives provided through other voluntary programmes.

As of June 2004, there were 34.8 million acres enrolled in CRP, with 2.8 million of those acres in partial-field enrolments under the continuous signup, Conservation Reserve Enhancement Program (CREP), and the Farmable Wetland Program (FWP) (USDA-FSA, 2004). More than 660 000 contracts were in force with more than 390 000 farmers covering this acreage. The annual rental cost was almost USD 1.7 billion, and the average rental cost per acre was USD 48. About 60% of CRP acreage was planted to grasses, 16% to trees or woody vegetation for wildlife, and 5% was dedicated to wetland restoration. The programme is authorized at 39.7 million acres for enrolments under 10-15-year contracts through 2007. The current CRP program targets land retirement to increase the cost-effectiveness of the programme, and adjusts the payment to closely match the market value.

Targeting

In the 1930s, when land retirement was first tried as a policy instrument, little attention was paid to specific criteria for identifying which land to retire. Quantitative tools developed for conservation planning at the farm level were extended to conservation policy planning at the regional and National level during debates about the 1985 omnibus farm legislation. A kind of "triage" was performed on the Nation's cropland (Bills and Heimlich, 1984). Land that would not result in high erosion rates even if farmed intensively was deemed "nonerodible". Land on which erosion rates could be successfully controlled through the use of conservation tillage and other practices, while remaining in crop production, was classed as "moderately erodible". "Highly erodible land (HEL)" was defined as land that, even if farmed under the most rigorous conservation practices, would still produce unacceptable rates of soil erosion.

CRP, enacted in the 1985 Food Security Act (FSA), adopted this targeting scheme, restricting enrolment to highly erodible cropland. A goal of enrolling 45 million acres of the

142 million acres of HEL by 1990 was included in the legislation. Of 36.4 million acres enrolled between 1986 and 1993, 8% were eroding at more than 40 tonnes per acre per year (TAY), and 83% were eroding at greater than 10 TAY (Osborn *et al.*, 1995). Later analysis showed that more than half of the land enrolled was HEL, resulting in savings of 281 million tonnes of soil per year (USDA-SCS, 1994). Average erosion rates on HEL land were reduced from 8.6 to 0.6 TAY for water-caused erosion, and from 10.7 to 1.3 TAY for wind erosion. In total, CRP enrolment reduced total erosion by 12%, accounting for 38% of erosion reductions between 1982 and 1992. Of 1.2 billion tonnes of sheet and rill and wind erosion reduced between 1982 and 1997, on net from all sources, 406 million tonnes (34%) was due to CRP (Claassen *et al.*, 2001).

The focus on HEL land implicitly recognized that soil erosion reduction was the primary environmental objective of the programme and enrolled land with the greatest inherent capacity to produce erosion. By 1990, however, it became clear that soil erosion itself was not as important a goal has had been previously thought. In particular, physical and economic studies of the effects of observed rates of erosion on crop productivity showed that onsite impacts paled in comparison with the impact of erosion and sedimentation on water quality, fish and wildlife habitat, and public services such as dams, ditches, and canals (Crosson and Stout, 1983; Larson *et al.*, 1983; Clark *et al.*, 1985; Ribaudo, 1986; AAEA, 1986). These findings sparked growing interest in conservation policies that would mitigate the offsite impacts of erosion, and related nutrient and pesticide runoff (Ogg *et al.*, 1989). In addition, the importance of land retirement for creating and improving wildife habitat associated with farmland, and the social and economic benefits people derived from these changes, prompted greater attention to these impacts (Berner, 1989; CAST, 1990; Allen, 1994). Finally, rebounding commodity prices blunted the desire to meet the 45 million acre enrolment goal, resulting in a 36 million acre enrolment cap in the 1990 FACTA act. Enrolling the "best" acres (those with the highest environmental benefits per dollar spent) with the remaining acreage became more important than enrolling as many acres as possible.

As a result of these three factors, following the 1990 Farm Bill USDA was encouraged by Congress to enrol lands in CRP that achieved the greatest environmental benefits for each dollar spent. In order to meet the multiple environmental objectives and to increase the cost-effectiveness of the programme, USDA developed and instituted an Environmental Benefits Index (EBI) designed to proxy for the range of environmental benefits being sought (USDA-FSA, 1997). The EBI was evaluated for every parcel of land offered, and included terms for:

- improvements in surface water quality;
- improvements in groundwater quality;
- maintenance of soil productivity;
- assistance to producers with potential problems implementing conservation compliance plans;
- acreage planted to trees;
- acreage within identified critical water quality problem areas; and
- acreage within conservation priority areas designated by Congress.

A national cost-effective ranking based on the EBI score and offered rental rate was constructed for each signup. Bids with the highest ratios were accepted until the acreage enrolment objectives for the signup were met. Analysis indicated that the post-1990 bid acceptance processes were targeting more expensive land in the Corn Belt and Northeast, with higher water-caused (sheet and rill) erosion rates, impacting water quality problems. The EBI has been updated several times, with the latest version for the 26th signup at USDA-FSA (2003).

A more formal validation of the new procedures was accomplished in an *ex post* evaluation of signups 1-12 and the new procedures (Feather *et al.*, 1999). They found that EBI criteria increased freshwater-based recreation and wildlife-viewing benefits, and decreased pheasant-hunting benefits compared with CRP acreage accepted prior to 1992. Based on this partial accounting of benefits, the new procedures increased the benefits USD 370 million per year.

In 1996, when the 10-year contracts originally made in 1986 started to expire, CRP was reauthorized, and use of the EBI to ensure cost-effective enrolment was confirmed.

Getting the rent right

Compensating the farm operator for the opportunity cost of not using the land in crop production is the economic basis for all US land retirement programmes. Without such compensation, and in the absence of any regulations against using this land for crop production, it is impossible to get farm operators in any numbers to offer land for enrolment. Despite this simple criterion for payment, determining the correct compensation for individual parcels is a difficult task for an operational programme covering millions of acres. In well-functioning, competitive markets, annual rents for agricultural land are theoretically equivalent to the annual returns from agricultural production, and hence equal the opportunity cost of using the land. This is not always the case, as discussed below.

In theory, auctions could promote efficiency in running such a programme (Dicks, 1985; Ervin and Mill, 1985). Government and the farm operator could engage in bid/acceptance behavior that would let farmers offer land at a bid price, and government accept those bids it desired. In fact, an experimental bid programme was conducted in 1958, but not continued (Christensen and Aines, 1962, p. 45). When the modern CRP was authorized in 1985, such a bid/acceptance process was implemented. However, the 45 million-acre goal effectively precluded auctions because it was impossible to enrol that much acreage in a short period. Multiple enrolment periods and the obvious pressure government officials were under to enrol as much land as quickly as possible opened opportunities to "game" any bidding system in favor of the landowner. Unprepared to negotiate bids with landowners, USDA officials sought guidance for Maximum Acceptable Rental Rates (MARR) based on county average rents that rapidly evolved into an offer system set at the MARR.

By 1990, the failure of the bid system and problems with the MARR became obvious. A soil-adjusted rental rate system was proposed that captured the essential elements of the distribution of soil productivity (Reichelderfer and Boggess, 1988; Barbarika *et al.*, 1994). Average county rents are adjusted up and down in relation to the ratio of the parcel soil productivity to average productivity in the county. For example, a parcel with soil that is 20% more productive than the "average" soil in the county would get 1.2 times the average county rent. This system adjusts the compensation paid to the single factor most likely to affect the opportunity cost of operating land, relative crop productivity.

Continuous signup

Another innovation in the 1996 Farm Bill was to allow partial field enrolment of land with certain high priority conservation practices, such as filter strips and riparian buffers, at any time during the year without competition. In addition to annual soil rental payment and cost-share assistance, many practices are eligible for additional annual and one-time up-front financial incentives. A related State-Federal cooperative programme called the Conservation Reserve Enhancement Program (CREP) focused continuous signup on locally-identified projects designed to address specific environmental objectives through targeted CRP enrolments. Sign-up is held on a continuous basis, but general sign-up practices may be included in addition to those under the regular continuous signup, and additional financial incentives are generally provided. There were 29 agreements in effect in 25 states as of 2004.

CRP, and land retirement programmes in general, have evolved from simple beginnings in the 1930s as new objectives and new administrative procedures were developed. The transaction cost data for the programme reflect these changes, especially in years initiating new 10-year contract periods. The analysis attempts to separate out these changes, to some degree.

Evolving from whole-farm planning to programme-based technical assistance

Prior to the 1985 Food Security Act, which established the Conservation Reserve Program and conservation compliance requirements, USDA's NRCS[2] relied on a whole-farm conservation planning process that identified all resource concerns on the farm.

"Conservation districts and SCS had championed the complete farm conservation plan approach [...] Most districts assigned priority to farmers who were interested in more than single-practice assistance" (Cohee, 1986, p. 95).

With limited cost-share funding (ACP was limited to USD 3 500 per farm per year), some elements of the plan might never be implemented. Many farmers surveyed in a conservation planning study said "SCS personnel often recommended practices that were too elaborate, not cost-effective, or too expensive, even with cost-sharing" (Nielson, 1986, p. 76). SCS tended to work with farmers who had been co-operators in the past, not necessarily those with the most critical conservation problems. District conservationists often found it difficult to interest new farmers in becoming co-operators because there were few financial incentives to do conservation planning and implementation.

Passage of the CRP and conservation compliance provisions created a demand pressure for conservation planning that nearly overwhelmed SCS in 1986. Even as the programme was being legislated, congressional staff worried that expanding eligibility on top of other changes in conservation programme changes "might overload the system" (Gray, 1986, p. 31). One observer noted "implementation of [CRP] will require substantial reallocation of staff time and expenditures by a variety of federal, state, and local government agencies [...] The opportunity costs of these reallocations will not be recognised until other programmes have been neglected" (Ervin and Blase, 1986, p. 80). Assisting landowners in determining whether land met the eligibility criteria for the nearly 101 million acres of HEL cropland and developing conservation cover plans for CRP after acceptance on more than 8 million acres signed up in 1986 alone caused SCS to re-evaluate its planning processes.

However, a survey of USDA county conservation officials found that only 10-12% of respondents thought scheduling or obtaining technical assistance was a problem limiting farmers' enrolment in CRP (Nowak and Schnepf, 1989, Table 7). The need for more personnel at the local level to implement the programme was mentioned more than any other consideration, however, by one of every five respondents who commented. Respondents also reported greater local interagency cooperation in implementing the programmes, and that level increased with increasing erosion rates and percentage of land with highly erodible soils. Local conservationists assessed farmers' paperwork burden in applying for CRP as an important barrier in 18-19% of responses. ASCS county officials and county Extension agents cited this as a barrier about twice as often as SCS District conservationists (Nowak and Schnepf, 1989, Table 7). These pressures on local staff have grown, with district and area conservationists recording increased scores toward "burnout" later in the Food Security Act implementation schedule (Barlow, 1989).

The pressure of limited time frames for both CRP eligibility assessment and conservation cover planning, and planning for conservation compliance led SCS to consider some radical changes to their traditional one-on-one, whole-farm conservation planning process. Conservationists narrowed their focus from all resource concerns on the farm to what farmers wanted to do (or had to do) with HEL cropland. Table 6.1 presents the findings of a 1989 SCS survey of alternative planning techniques.

Table 6.1. **Cost of conservation plans by planning process**

Planning process	Source of assistance	Districts	Number of plans	Acres planned	Cost per plan	Cost per acre		
						Average	Minimum	Maximum
		Number	Number	Acres	USD	USD per acre planned		
One-on-one	Contractor	2	1 668	697 150	91.10	0.22	0.18	0.22
One-on-one	SCS sole source	5	4 694	912 968	182.87	0.94	0.76	2.56
One-on-one	SCS in lead role	10	10 075	1 548 917	154.99	1.01	0.38	3.62
Group	SCS sole source	1	307	19 700	73.09	1.14	n.a.	n.a.
Combination	SCS in lead role	11	19 259	2 807 413	197.21	1.35	0.20	4.01

n.a.: not available; SCS: Soil Conservation Service.
Source: Robertson, *et al.* (1989).

However, the complexity of soils, landscapes, and practices probably overwhelmed any differences in cost because of the planning process itself. The authors pointed out that group processes would be more effective when soil types, cropping patterns, and conservation needs were similar, such as when planning similar cover types for similar highly erodible soils in a relatively homogeneous area. Other agency officials involved in conservation planning also found that group planning processes could meet conservation planning needs within the limited time frames, if proper materials and background were prepared in advance (Farnsworth and Braden, 1988; Farnsworth *et al*, 1988). Adoption of office automation and Geographical Information Systems (GIS) technology to facilitate the planning process added its own stresses to local planners who were not "computer literate" (Ventura and Giampetroni, 1992).

The agency began to realize that the narrow focus on planning for erosion control on highly erodible land in CRP implementation failed to capitalize on potential opportunities to plan for water quality, wildlife habitat, and other resource concerns which moved to higher priority as the 1996 Farm Bill was being considered (Bridge, 1993). New programmes established in the 1996 FAIR Act, along with reauthorisation of CRP put even more money, and therefore more pressure, on the conservation establishment to eschew whole-farm planning in favor of planning just enough to get the farmer's application for one of many programmes (CRP, EQIP, WHIP, FRPP, etc.) completed. Each of these programmes has their own requirements for planning and applications, so the local conservationist's time is even more fragmented than ever before.

6.3. Interagency roles in the Conservation Reserve Program

There has been a long-standing division of labor within USDA in implementing agricultural conservation programmes. Responsibility for educating and providing technical assistance to farmers rests with NRCS, while financial assistance paid to farmers was administered through FSA. These are the agencies with principal responsibilities and funding for conservation programmes directed toward production agriculture. Other USDA, Federal, State and local government agencies have also played roles in developing and implementing conservation programmes, but their participation is usually short-lived, *ad hoc*, a minor financial commitment, and not easily captured in agency budget accounts.

In CRP, NRCS responsibilities include certifying that land offered meets eligibility requirements for soil erodibility and other criteria, assisting with quantification of the EBI, working with the producer to develop a conservation cover plan for the land, providing technical assistance and oversight in establishing the conservation cover, and monitoring that cover is successfully established and maintained. Thus, much of the technical assistance provided by NRCS occurs at the beginning of the contract, tapering off rapidly once conservation cover is successfully established. NRCS staff handling CRP do so on a periodic basis, and then move on to other programmes and responsibilities. On the other hand, FSA responsibilities include verifying that the producer is financially eligible to participate, assisting the producer in submitting a signup application, submitting the EBI score, notifying successful applicants and finalizing the contract, administering payment, and verifying that contracts remain eligible for payment. These responsibilities continue at a fairly high level over the life of the contract. While some FSA staff in county offices may be drafted to work on CRP general signups during short periods (which is probably not captured in reported FSA expenditure data), many are assigned full time to service CRP. Specific roles in CRP at the State and local office level are detailed in Table 6.2.

NRCS technical assistance prior to acceptance thus focuses on the CRP-2 form (see Figure 6.2), certifying the eligibility of the land, the suitability of the proposed conservation cover practices, and the soil type (to which the soil rental rate is tied). FSA administrative support focuses on the CRP-1 form, which establishes the soil rental rate and certifies that crop insurance requirements are met.

FSA, primarily in the Washington office, runs the CRP bid acceptance process, which is largely invisible to the producer or landowner. Once a bid is accepted, administrative support and technical assistance again devolve to the local conservation district office at the county level. The producer is notified and NRCS (sometimes with USDA-Forest Service or USDOI-Fish and Wildlife Service assistance) helps develop a conservation plan of operation to install the conservation cover practices. FSA approves the plan of operation and, as it is implemented, issues CRP-1 contracts covering the various practices and ensuring processing for payment of cost-share and rental payments. NRCS district conservationists continue to follow establishment and maintenance of conservation cover on an as-needed basis throughout the life of the contract.

A detailed listing of roles for each major agency is contained in Table 6.2.

Table 6.2. **Roles of FSA, NRCS and FS county officials in Conservation Reserve Program administration and technical assistance**

Official or designee	Role
FSA District Director	Reviews and approves CRP-1s for USDA and related county level employees
	Ensures that an environmental evaluation (EE) has been completed for each CRP contract and all necessary consultations are complete
FSA County Executive Director (CED)	Computes cost-share and rental payments
	Manages day-to-day activities of field service centres and employees
County Office of the FSA	Calculates Maximum Annual Rental Rate (MARR)
	Determines producer and cropping history eligibility
	Determines if crop insurance requirements are met
	Completes "paid-for" measurement
	Reviews 25% enrolment limit for County
	Ensures CRP and AMTA contract acreage does not exceed agricultural use acreage on farm
	Determines reductions in quotas and allotments
	Makes the rental, cost-share (C/S), SIP, and PIP payments
County Committee (COC)	Approves eligible cropland acres and conservation plans
	Can delegate some of its authority to CED (*e.g.* eligible cropland acres and approval of conservation plans)
	Approves CRP-1 except for USDA, FSA, Conservation District, and Headquarters' office employees and members
	Determines violations of CRP-1s
	Provides written approval to landowners, at the end of the easement, to have the easement removed from title
	Approves share agreements of owner/operator
	Can provide certain ownership eligibility waivers
	Determines compliance with landlord/tenant provisions on participation
	Provides guidance to county office, determines producer, land and practice eligibility for CCRP
	Ensures conservation plan for CCRP includes maintenance practices
	Approves final conservation plan and CRP-1 for CCRP
	After concurring with the state FSA committee, they can decide to hold
	continuous sign-up only if 25% county limit is not reached or can ask for a waiver from the state FSA committee
	Determines if the annual payment limit of USD 50 000 is exceeded and then reduces payment
	Authorizes most cost-share agreements
	Can establish cost-share rates if authorised by state FSA committee and with NRCS concurrence
District Conservationist-NRCS Field Office	Represents NRCS with FSA, COC, State Forestry and State wildlife agencies, Conservation Districts
	Coordinates tree planting with State Forestry
	Determines practice suitability, need, and feasibility of practice and predominant soil types for determining land eligibility
Conservation District	Approves tree planting plan
	Provides letter of recommendation to COC to exceed 25% enrolment, if need be
	Approves conservation plans
NRCS	Participates in state level technical determinations and policy reviews
	Determines EBI scores for factors #1 through #6 and fills in on CRP-2
	Assist county offices in identifying soil types
	Develops conservation plan and cost-share agreement with FS if applicable
	Completes site specific EE
	Performs annual status review
	Obtains conservation district approval of conservation plans
Forest Service (FS)	Develops tree planting plans
	Provides technical assistance for tree planting practices
	Monitors and certifies practice compliance
	Develops stewardship plans for converted CRP land

Source: Table 3.2-1, USDA, FSA Conservation Reserve Program, Final Programmatic Environmental Impact Statement.

Figure 6.2. **Diagram of general CRP contract process**

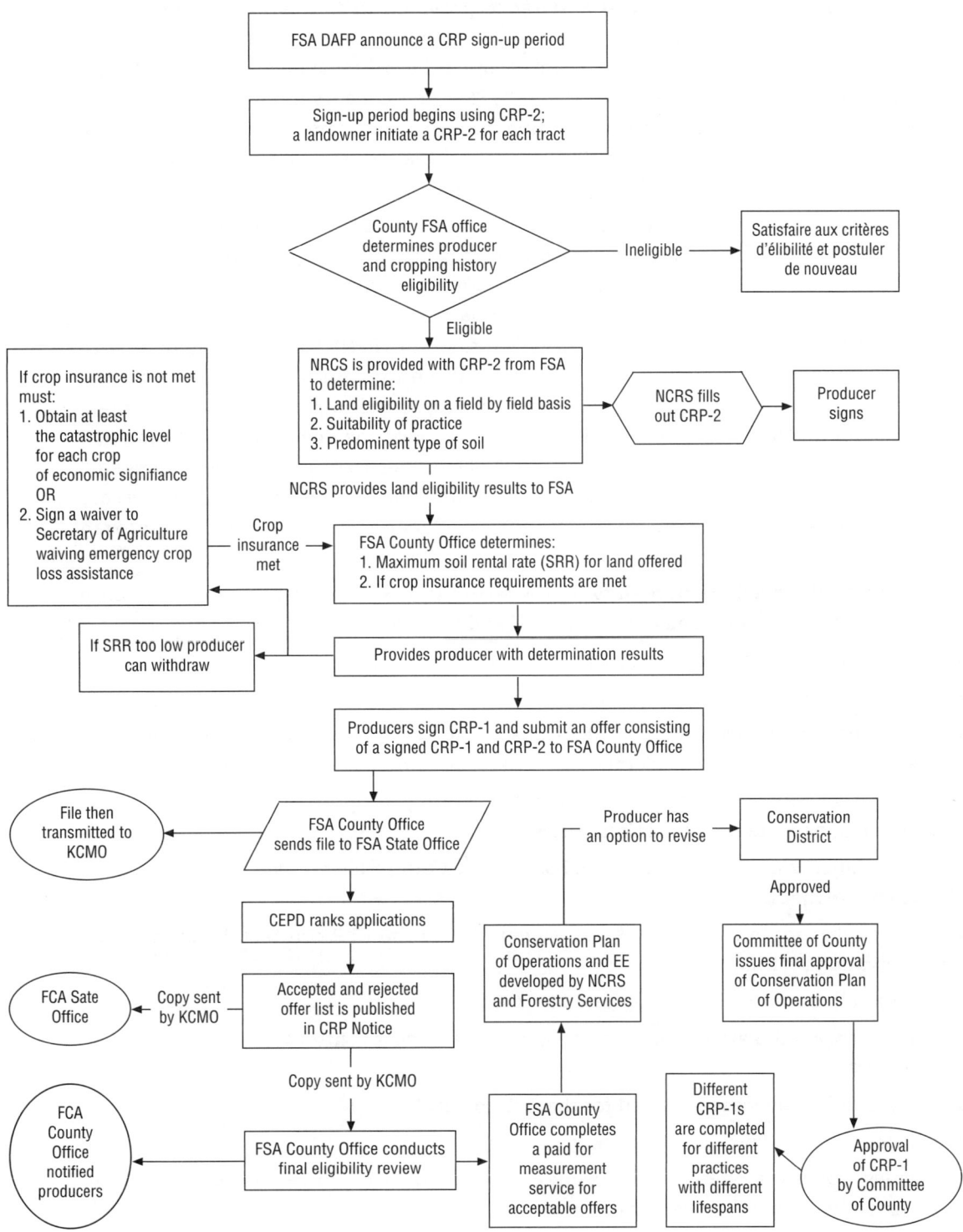

KMCO: Kansas City Management Office.

Source: USDA, FSA, Conservation Reserve Program, Final Programmatic Environmental Impact Statement.

Conservation planning

The primary form of technical assistance provided in CRP is preparation of the conservation plan for CRP acres. An approved conservation plan is required before a CRP contract can be approved. The plan is a record of supporting information and decisions for the treatment of the CRP parcel. Prior to the 1985 FSA Act, conservation plans were often comprehensive, covering the entire farm operation. For CRP (and many other financial assistance programmes passed since 1985), a programme conservation plan only needs to contain information related specifically to the CRP parcel.

The participants involved in the development of the CRP conservation plan include NRCS District Conservationist, State forester (if trees are involved), FWS or State wildlife planner (if habitat is involved) and FSA's county committee. NRCS is ultimately responsible for the technical leadership for planning and implementation, adherence to NRCS policy in the National Planning and Procedures Handbook (NPPH) about compliance with the US National Environmental Policy Act (NEPA), and technical concurrence on the conservation plans and any revisions. However, FSA is the lead agency with ultimate responsibility for NEPA compliance. NRCS would complete the Environmental Evaluation for any potential threatened and endangered (T&E) species issues under the Endangered Species Act in conjunction with required field visits. FSA then ensures that necessary consultations are carried out. Ultimate approval depends on whether the approved conservation plan:

- contains all the practices necessary for the successful establishment and maintenance of the vegetative cover on all of the acres offered for CRP;
- is technically adequate for achieving CRP objectives;
- adheres to NRCS policy about NEPA compliance;
- is reviewed and approved by the NRCS Conservation District (the district may assist the producer in planning and implementing conservation management systems); and
- ensures that CRP cover will not be disturbed during the primary nesting season, as determined by State Technical Committee.

Before approving CRP contracts, the FSA County Committee reviews and approves the plan to ensure that the plan:

- has been signed and agreed to by all signatories to the CRP contract, NRCS, and the Conservation District;
- includes all of the eligible acres offered for CRP;
- includes required maintenance for weed, insect, and pest control for the life of the CRP contract;
- includes only practices requested for the CRP contract;
- includes C/S for eligible practices only;
- includes application rates, such as the amount of seed, lime, fertiliser, that are consistent with practice specifications; and
- meets the objective of the conservation priority area (CPA), if enrolled in one.

The approved conservation plan obligates CRP participants to establish and maintain approved practices that:

- where appropriate, plant perennial seeding and planting mixes that achieve the highest environmental benefits for each CRP practice;

- where practical, use State-certified seed for CRP (common seeds, especially natives, may be used when certified seed is not available);

- where appropriate, avoid the use of single, introduced species;

- use native legumes, forbs, shrubs, and plant mixes; and

- ensure that the approved seeding mix does not include weed species, including noxious weeds.

The most important aspect of a CRP conservation plan is that it outlines the necessary maintenance practices for the successful establishment and maintenance of the approved practices included in the CRP contract, regardless of the applicant's eligibility for cost-share funds.

US Fish and Wildlife Service, Partners for Fish and Wildlife Program

Fish and Wildlife Service biologists provide biological expertise on habitat management, restoration and individual species needs on lands enrolled in the CRP, EQIP, WHIP and a variety of similar conservation programmes. Since 1992, the Partners Program has helped NRCS and landowners choose sites for the WRP, craft restoration plans, and participate in decisions on land use activities inside the WRP easement area.

In recent years (2001-04), about USD 1.2 million per year in FWS technical assistance has been provided to CRP applicants and contract holders (Naley, 2004). This figure was likely larger in start-up years when many more acres and contracts were being applied for and conservation cover established.

6.4. CRP technical assistance and support costs

Because responsibility for CRP is split between FSA and NRCS (and to a limited extent USDA's Forest Service), USDA transaction costs must include both NRCS and FS technical assistance costs and FSA administrative support costs. Costs are estimated using aggregate budget data. The limitations of such data are explained in Box 6.1. Since the USDA budget detail does not break out FSA conservation support by programme, the total is apportioned by the relative expenditures on the CRP (by far the largest), ACP, Emergency Conservation Program (ECP), and Rural Clean Water Program (RCWP) in each year. The resulting amounts, expressed as a percentage of rental and cost-share payments for cover establishment, are shown in Figure 6.3.

Both NRCS/FS and FSA had substantial start-up costs (87% for NRCS/FS and 23% for FSA) reflected in the amount of technical assistance and support expenditures per dollar of payments in 1986, the first year CRP was operated. However, FSA support expenditures settled down to a steady 3-4% of payments, while NRCS/FS technical assistance costs were far more variable, ranging from 0.4 to 4% of payments. This reflects the steady nature of administrative support provided by FSA for contracts, and the episodic nature of technical assistance effort required to plan for and implement cover establishment for land entering CRP at different times. Technical assistance costs receded as enrolment proceeded, fully enrolling the 36 million acres allotted to the programme. After CRP was reauthorised in the

---

Box 6.1. **A note on data quality**

The data used in this analysis is aggregate budget data reported through the US Department of Agriculture. Any such administrative data has a host of problems associated with budget processes and administrative accounting rules. In addition to the usual problems, the data on NRCS technical assistance costs for CRP suffer from two additional problems.

First, district conservationists in the field are under severe pressure to both balance the demands from producers and clients and also account for their time within the funding codes available to them in the accounting system. Incentives are on the side of making sure that both clients and accounts are balanced on paper, whether they are in fact or not. Consequently, some work reported as "CRP technical assistance" may not be, and *vice versa*.

Second, NRCS performs CRP technical assistance under reimbursement from FSA rather than through direct appropriation. There is an understandable bureaucratic tendency to attempt reimbursement for as much as work as possible, perhaps covering other work not strictly relating to CRP. In addition, the controversy over the Section 11 CCC funding cap may have limited the amount of reimbursement NRCS sought. The net effect of potential over- and under-reporting of CRP technical assistance costs is impossible to determine from these data.

Despite these flaws, these data are the official budget costs reported by USDA and the only practicable source for estimating transaction costs at the national level.

Figure 6.3. **Conservation Reserve Program: Technical assistance and support as a per cent of cost-share and rental payments**

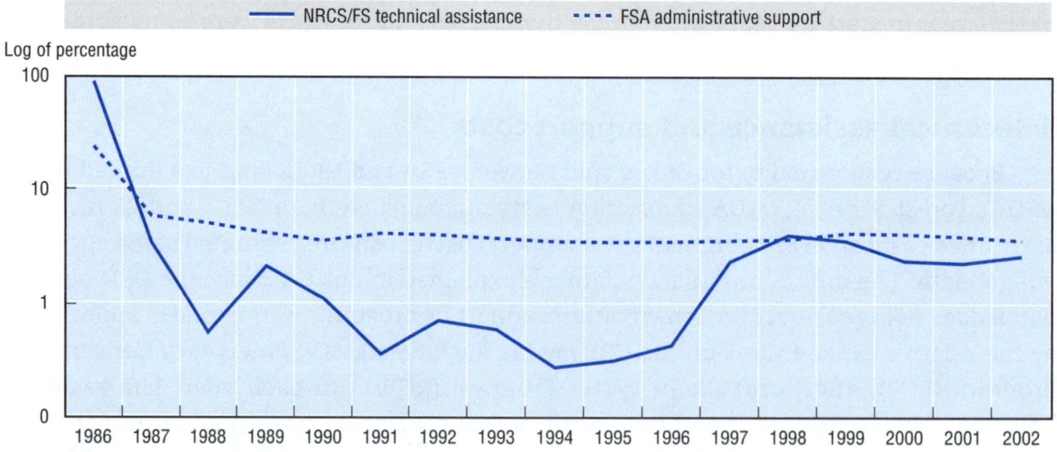

Source: USDA, OBPA and Ralph E. Heimlich, Agricultural Conservation Economics.

1996 FAIR Act and previously enrolled acreage came out of contract and was resubmitted, NRCS/FS technical assistance costs rose again. The addition of CREP and institutionalisation of the continuous CRP signup provided a steadier, ongoing technical assistance demand reflected in the higher level of continuing assistance after 1996.

A better metric for transaction costs than dollar expenditures is costs per enrolled acre and cumulative enrolled acre (Figure 6.4). Because FSA administrative support extends to both newly enrolled and continuing acres, the costs per newly enrolled acre climb steadily (no acres were enrolled in 1994 and 1995), while costs per cumulative acre flatten out and

Figure 6.4. **Conservation Reserve Program: Transaction costs per new and cumulative acre enrolled**

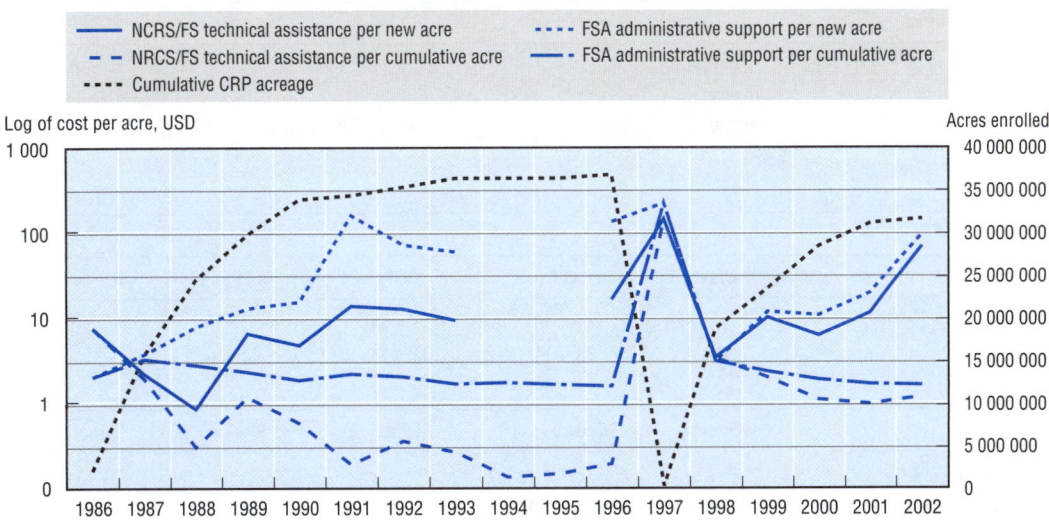

Source: USDA, OBPA and Ralph E. Heimlich, Agricultural Conservation Economics

then recede. Costs per newly enrolled acre are less than proportional to enrolment after the initial burst, probably because personnel learn their roles and become more efficient in delivering the programme for additional acres. Participants also may learn, after repeated attempts to enrol, requiring less administrative and technical help.

Technical assistance for signup and cover establishment by NRCS and FS follows a similar pattern of rise as signups increase, with declines in cost per cumulative acre enrolled, but with more variation. Because technical assistance needs fall off after the first few years of a contract, expenditures per cumulative acre fall off more dramatically than for FSA administrative support.

Major bursts of administrative and technical assistance activity occurred in 1986, when the programme was being developed and deployed, and again in 1997. Both FSA and NRCS/FS expenses expanded in 1996-97 when the second round of enrolments was being prepared after CRP was reauthorized in the 1996 FAIR Act, and when the largest amount of acreage from contracts expiring and being re-enrolled was processed for retirement beginning in 1998. Once again, costs per cumulative acre fall off, with technical assistance costs dropping faster than administrative support.

As time series of transaction costs for CRP are available, it is possible to decompose how various factors affect transaction costs. A simple linear regression equation relating transaction costs to summary characteristics for the year in which the signup occurred is revealing. In order to account for the differences between administrative support and technical assistance, separate estimations were made for each.

The equation for FSA administrative support (Table 6.3) is highly predictive, explaining nearly 80% (adjusted R^2 = 0.776) of the variation in costs over the 17 years estimated, with a significant value for the F test. Four variables are statistically significant in the equation. Cumulative acreage enrolled has the most significance, which is consistent with the idea that administrative support is spread across all the acres in the programme, not just those newly enrolled (Figure 6.5). The coefficients are interpreted as marginal costs, so each acre added to the programme is estimated to account for USD 1.79 in FSA administrative

Table 6.3. **Regression equation of Conservation Reserve Program FSA administrative support expenditures, 1986-2002**

Regression statistics					
Multiple R	0.961				
R Square	0.923				
Adjusted R Square	0.776				
Standard Error	5 496 547				
Observations	17				

ANOVA

	Df	SS	MS	F	Significance F
Regression	7	4.E + 15	5.E + 14	17	0
Residual	10	3.E + 14	3.E + 13		
Total	17	4.E + 15			

	Coefficients	Standard error	t Stat	P-value	Lower 95%
Acres newly enrolled	−6.56	2.16	***(3.03)	0.01	(11.37)
Acres idled/installed	−0.20	0.45	(0.43)	0.67	(1.21)
Reenrolled acres	0.68	0.64	1.08	0.31	(0.73)
Continuous acres	−3.83	9.17	(0.42)	0.69	(24.25)
Cumulative acres enrolled	1.79	0.08	***22.62	0.00	1.62
Number of contracts enrolled	798.35	238.32	***3.35	0.01	267.33
Post1996 dummy	−41 196 527	10 543 697	***(3.91)	0.00	(64 689 352)

*** significant at the 95% confidence level.
Source: Ralph E. Heimlich, Agricultural Conservation Economics.

Figure 6.5. **Conservation Reserve Program:**
Actual and simulated FSA administrative support costs

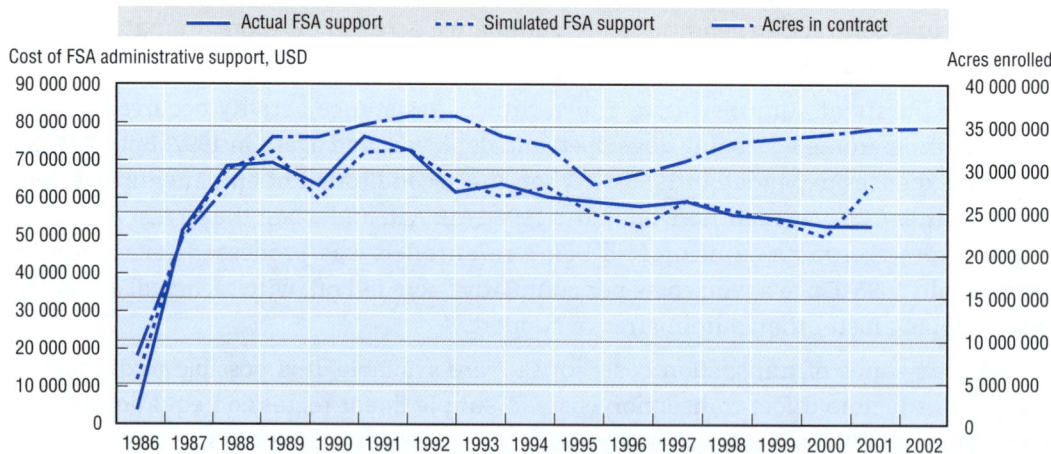

Source: USDA, OBPA and Ralph E. Heimlich, Agricultural Conservation Economics.

support costs. The number of contracts enrolled is significant, each additional contract adding USD 798 to costs. This probably reflects some economies of scale in preparing for signups, and some efficiency from "learning-by-doing" as more contracts are handled. The acres enrolled in each signup is significant, with each additional acre lowering costs by USD 6.56. This is logical considering that larger contracts spread the costs of contract administration over more acres and reduce the total number of contracts that must be administered. Finally, a dummy variable accounting for the difference between the

1985-95 programme and the 1996-2002 programme is statistically significant, shifting FSA support costs down by USD 41.2 million per year. This may reflect a reduced burden on the FSA field staff because of streamlining the bid selection process, standardising the rental rates, and automating much of the signup paperwork through web-based applications.

Other coefficients, although not statistically significant, are interesting for the direction and magnitude of their estimated effect on administrative support costs. Each acre idled subtracts USD 0.20, presumably because of the effect of spreading fixed costs over more acres. In later signups (after 1996), land previously enrolled in CRP could be re-enrolled. These are estimated to add USD 0.68 per re-enrolled acre. Land brought into the programme under the continuous signup process incurs considerably less administrative costs than land enrolled in the general signup. These are estimated to subtract USD 3.83 from administrative support costs per acre enrolled in the continuous programme. Note that this only pertains to administrative transaction costs: rental and cost-share payments for continuous acres are often higher than for general signup acreage.

There is considerably more variation in NRCS/FS technical assistance costs, and because they are generally incurred in the first contract years, different variables are expected to explain that variation (Table 6.4). Only about half of the variance in the data is explained by the regression (adjusted $R^2 = 0.472$), and the regression F test is not as significant as the FSA cost equation. As expected, cumulative acres enrolled is not as significant a variable in this equation as for FSA costs, and the coefficient on cumulative acres is much smaller, adding only USD 0.30 per acre enrolled. Significant explanatory variables for technical assistance costs are the number of acres idled or for which cover

Table 6.4. **Regression equation of Conservation Reserve Program NRCS/FS technical assistance expenditures, 1986-2002**

Regression statistics					
Multiple R	0.856				
R Square	0.732				
Adjusted R Square	0.472				
Standard Error	11 432 253				
Observations	17				

ANOVA					
	Df	SS	MS	F	Significance F
Regression	7	4.E + 15	5.E + 14	4	0
Residual	10	1.E + 15	1.E + 14		
Total	17	5.E + 15			

	Coefficients	Standard error	Stat	P-value	Lower 95%
Acres newly enrolled	1.40	4.50	0.31	0.76	(8.62)
Acres idled/installed	2.39	0.94	***2.54	0.03	0.29
Reenrolled acres	−0.51	1.32	(0.39)	0.71	(3.46)
Continuous acres	−3.63	19.07	(0.19)	0.85	(46.11)
Cumulative acres enrolled	0.30	0.16	**1.84	0.10	(0.06)
Number of contracts enrolled	−220.82	495.69	(0.45)	0.67	(1 325.28)
Post 1996 dummy	31 894 665	21 929 806	*1.45	0.18	(16 967 998)

* significant at the 80% confidence level.
** significant at the 90% level.
*** significant at the 95% level.
Source: Ralph E. Heimlich, Agricultural Conservation Economics.

was installed in each year number of acres enrolled in each year (not cumulative), and the dummy for post-1996 enrolment. Not surprisingly, each additional acre idled/installed adds USD 2.39 (Figure 6.6). The post-1996 dummy variable indicates that NRCS technical assistance costs shifted up by USD 31.9 million per year after 1996. This may be explained by the additional complexity involved in evaluating the EBI, which was initiated in 1991, but wasn't used on a large acreage until after the 1996 CRP reauthorisation. The increase may also reflect the more complex technical assistance for additional environmental issues dealt with by the programme, such as wildlife habitat and water quality issues, compared with technical assistance for soil erosion on HEL in earlier signups.

Figure 6.6. **Conservation Reserve Program:**
Actual and simulated NRCS/FS technical assistance expenditures

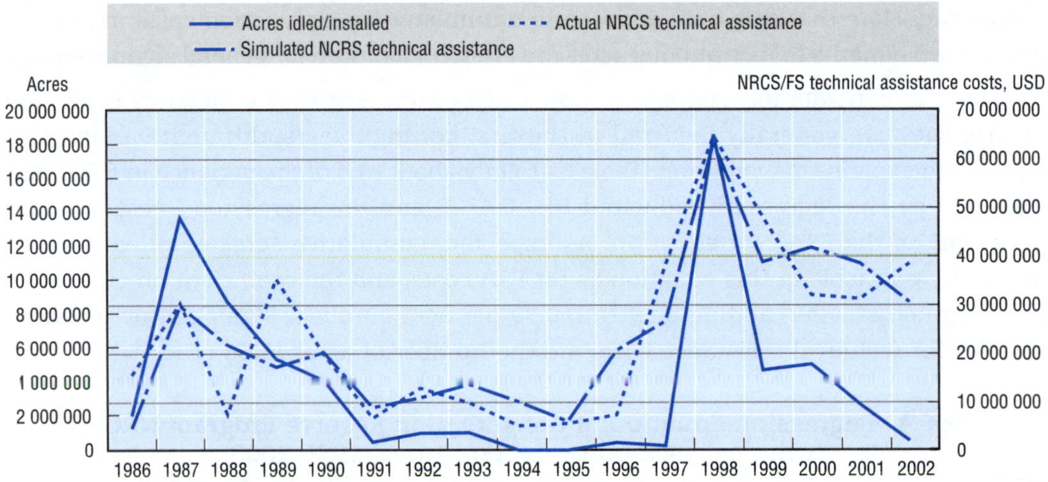

Source: USDA, OBPA and Ralph E. Heimlich, Agricultural Conservation Economics.

Although not statistically significant, the coefficients on remaining variables are interesting in sign and magnitude. Reenrolled acres reduce technical assistance costs by USD 0.51 per acre. Acres enrolled in the continuous signup reduce technical assistance costs by USD 3.63 per acre, perhaps because these primarily focused on riparian buffer practices.

Technical assistance in initial and succeeding years

Evidence from expenditure data shows that there are substantial transaction costs in developing and rolling out conservation programmes in their initial years, relative to ongoing costs once programmes are established (Table 6.5). CRP's initial years were 1986, when the first enrolments in the programme were made, and 1997, when the first contracts under the programme reauthorized after the 1996 FAIR Act were enrolled. NRCS technical assistance costs averaged USD 0.03 per dollar of expenditures (3%) in the initial period, but only USD 0.01 per dollar (1%) in succeeding years. FSA administrative support costs were essentially equal per dollar of expenditure in both periods. Another measure is the cost per acre enrolled, which was USD 23.21 per acre for NRCS technical assistance in the initial years, and dropped to only USD 5.33 per acre enrolled in succeeding years. FSA administrative costs per acre enrolled also dropped, from USD 27.11 per acre to only USD 13.97 per acre.

Table 6.5. **Technical assistance and administrative support in initial and succeeding years of US conservation programmes, 1983-2002 (USD)**

		Conservation Reserve Program (CRP)	Wetland Reserve Program (WRP)	Environmental Quality Improvement Program (EQIP) and predecessors
	Initial year(s)	1986, 1997	1993	1995-96
Initial year(s), million 1996 constant dollars	NRCS technical assistance	53.4	5.3	194.3
	FSA administrative support	62.4	n.a.	10.7
Succeeding year(s), million 1996 constant dollars	NRCS technical assistance	353.2	85.5	1 476.5
	FSA administrative support	925.3	n.a.	168.5
Per dollar of expenditure, initial year(s), dollars per 1996 constant dollar	NRCS technical assistance	0.03	1.11	0.62
	FSA administrative support	0.04	n.a.	0.03
Per dollar of expenditure, succeeding year(s), dollars per 1996 constant dollar	NRCS technical assistance	0.01	0.09	0.37
	FSA administrative support	0.04	n.a.	0.04
Per acre enrolled, initial year(s), 1996 constant dollars per acre	NRCS technical assistance	23.21	106.93	n.a.
	FSA administrative support	27.11	n.a.	n.a.
Per acre enrolled, succeeding year(s), 1996 constant dollars per acre	NRCS technical assistance	5.33	93.38	n.a.
	FSA administrative support	13.97	n.a.	n.a.

n.a.: not available.
Source: USDA, OBPA and Ralph E. Heimlich, Agricultural Conservation Economics.

The publicly available administrative cost data used in this analysis do not discriminate between establishment costs, which might be considered an investment that should be amortized over the 10-year life of the CRP program, and ongoing costs associated with simply implementing the programme over its life span. While major programme design and redesign activities were undertaken in 1986 when modern CRP was first authorized, and again in 1996-97 when the programme was reauthorised, these were not the only times the programme was changed. For example, the EBI and soil-adjusted rental rates were developed after the 1990 FACT Act redirected the programme toward a broader array of environmental objectives, but these were mostly experiments used to meet the remaining 1.4 million acres under the 36.4 million acres enrolment cap. These methods were re-evaluated and refined after the 1996 FAIR Act reauthorised CRP for another set of 10-year enrolments and subsequently used to reenrol more than 22 million acres in 1997-98. There are other, more minor programme changes and administrative modifications made with every signup.

The contrast between establishment and ongoing costs is more marked for WRP (Table 6.5). NRCS technical assistance costs were more than USD 1.11 per dollar of expenditures in the initial year of the programme, and USD 106.93 per acre enrolled. After establishment, costs dropped to only USD 0.09 per dollar of expenditure (9%), and USD 93.38 per acre enrolled. WRP costs are considerable because of the need to establish a legal easement on the area to be restored as wetland, *versus* a simple contract between the producer and the government in other programmes. EQIP passed in the 1996 FAIR Act, consolidated several previous cost-share programmes, including the long-standing Agricultural Conservation Program (ACP), run by FSA, and the Great Plains Conservation Program (GPCP) and Colorado Salinity Control Programs, run by NRCS. The Forestry Incentives Program (FIP) was moved from NRCS to the Forest Service in 1996. Comparing technical assistance and administrative costs before and after the 1995-96 transition period shows that NRCS technical assistance costs dropped from USD 0.62 per dollar expended as EQIP was being established to the pre- and post-establishment average of USD 0.37 per dollar spent. FSA administrative costs remained constant at about USD 0.04 per dollar.

Comparing average annual costs for the first CRP program (1985-95) to those of the second programme (1996-2002) shows that NRCS technical assistance costs increased dramatically, while FSA administrative support costs fell (Table 6.6). NRCS costs increased 150%, from about USD 15 million per year to USD 36 million per year. Meanwhile, FSA costs dropped 6%. Overall costs increased 25%.

Table 6.6. **Differences in average annual agency transaction costs, first and second CRP (USD)**

	NRCS Technical assistance costs per year	FSA administrative support costs per year	Total agency transaction costs per year
First CRP (1986-96)	14 760 049	59 364 895	74 124 945
Second CRP (1996-2002)	36 996 689	56 289 935	93 286 624

Source: USDA, OBPA and Ralph E. Heimlich, Agricultural Conservation Economics.

There are probably at least five influences at work in these figures. First, use of the EBI in bid assessment and standardised soil-adjusted rental rates may have increased assessment costs for NRCS, but decreased costs for FSA. Even though these changes were first introduced in 1991, they were not applied to significant acreages until CRP was reauthorised in 1996 and the first set of contract began to expire. Second, the broader range of environmental issues dealt with after 1990, and particularly after 1996, probably required more technical assistance effort for NRCS in planning and implementing CRP cover practices. Third, NRCS may have become increasingly careful in accounting for technical assistance on CRP after 1996 because of the increasingly competitive demands on staff resources with the increase in conservation programme funding (but not staffing) in the 1996 FAIR Act. The continuous signup and CREP were implemented after 1996, which engendered a set of different technical and administrative efforts. Finally, FSA increasingly turned to web-based and GIS-enabled administrative tools for managing the CRP signups after 1996, which could have produced some savings.

Complex evaluation, open enrolment and cost-effectiveness

Superficially, the administrative and technical costs associated with the continuous and CREP enrolments might be thought to be lower than the work required to apply and evaluate the EBI under the general CRP signup. There is no direct data on transaction costs for the different enrolment methods, but the regression analysis of expenditures provides some limited support for this idea.

For FSA administrative support costs, land brought into the programme under the continuous signup process incurs considerably less administrative costs than land enrolled in the general signup. These are estimated to subtract USD 3.83 from administrative support costs per acre enrolled in the continuous programme, although the coefficients are not statistically significant (Table 6.3). NRCS and FS technical assistance costs for acres enrolled in the continuous signup are estimated to be USD 3.63 per acre less than for the general signup. Again, this coefficient is not statistically significant. Note that these estimates only pertain to administrative and technical assistance transaction costs: rental and cost-share payments for continuous acres are often higher than for general signup acreage. Average CRP rents in the general signup are USD 44 per acre, while they average USD 89 for continuous signup and USD 121 per acre for CREP (USDA, FSA, 2004).

While the riparian buffers, filter strips, and vegetative corridors accepted in continuous and CREP enrolments provide important environmental benefits for water quality improvement and wildlife habitat, it is not feasible to evaluate them against larger parcels accepted in the general signup. The smaller acreage involved means that fixed administrative and technical costs are amortised over fewer acres per contract, and the transaction costs for producers to enrol these smaller acreages are also larger. This is one reason that rental rates for these enrolments, as well as signing incentive payments and practice incentive payments, were raised to provide sufficient incentives to landowners. Despite these higher incentives, enrolment in continuous and CREP has lagged expectations. As of January 2004, only 555 626 CREP acres were enrolled of the 1.5 million acres allocated to State programmes (USDA, FSA, 2004b).

Continuous and CREP enrolment are complements to the larger whole-field enrolment of the general signup. While it may be useful to consider additional forms of continuous signup for other high-priority environmental practices on a partial-field, or even a whole-field, basis, it would not be cost-effective to replace the general signup with complete reliance on open enrolment.

6.5. Transaction costs for different kinds of conservation programmes

There are many ways to help farmers adopt conservation and environmental practices. A taxonomy of conservation approaches shows a continuum from the regulatory (rarely used in the US), through voluntary participation in response to financial incentives, to what is essentially moral suasion facilitated by education and technical assistance (Table 6.7). Transaction costs for these different approaches are quite different because the role of technical assistance varies from nearly absent (regulatory approaches) to dominant (programmes providing only technical assistance and education). For those programmes providing financial assistance for conservation, transaction costs for land retirement and cost-sharing/incentive programmes differ because of the timing, duration, and relative cost of technical assistance.

Technical assistance for cost-sharing and incentive programmes typically occurs in the same year in which the expenditure is made, and most installations are completed within a year or two. There is typically little need for follow-up technical assistance after the installation is complete. Technical assistance may be a much larger percentage of total costs, especially for practices that primarily affect how agricultural resources are managed, *versus* actual investments in machinery, structures, or materials. By contrast, technical assistance needs for land retirement occur in the first year or two to get conservation cover established, but payments go on for a period of years. Technical assistance is usually a much smaller percentage of total costs because rental and easement payments on land are typically quite large, and cover establishment costs are usually modest.

In either case, there may be extraordinary costs associated with developing a programme when it is first implemented and when it is being terminated. Development costs include working out technical standards, forms, staff training, reporting forms, procedures and data protocols, and financial accounting systems. Termination costs include finishing outstanding projects, resolving payment issues, finalising records and accounts, and transitioning staff to new duties.

Table 6.7. **Matrix of agricultural conservation/environmental problems, policy instruments, and federal programmes**

Conservation/environmental problems	Involuntary participation		Voluntary participation				Facilitative
Environmental problem	Regulation	Conservation compliance	Land retirement	Cost sharing	Incentive payments	Trading/banking/bonding	Education/technical assistance
Erosion: soil productivity loss		Sodbuster/compliance (1985)	Soil Bank (1956) CRP (1985)	ACP (1936)			CTA (1936)
Erosion: sedimentation	CZARA (1990)	Sodbuster/compliance (1990)	CRP (1990)	ACF (1936) EQIP (1996)	WQIP (1990) EQIP (1996)		CTA (1936)
Erosion: airborne dust		Sodbuster/compliance (1990)	CRP (1996)	ACF (1936) EQIP (1996)	ACP (1936) EQIP (1996)		CTA (1936)
Wetlands loss	CWA Section 404 (1972)	Swampbuster (1985)	Water Bank (1972) CRP (1988) WRP (1990) EWRP (1993)			Mitigation banking (1995)	
Water quality: impairment from nutrients	CZARA (1990)		CRP (1996)	EQIP (1996)	WQIP (1990) EQIP (1996)	CWA (1990)	CTA (1936)
Water quality: impairment from pesticides	FIFRA (1947) CZARA (1990)		CRP (1996)	EQIP (1996)	WQIP (1990) EQIP (1996)		CTA (1936)
Wildlife habitat loss	ESA (1973)		CRP (1996)	WHIP (1996)			

Acronyms: ACP – Agricultural Conservation Program. CRP – Conservation Reserve Program. CTA – Conservation Technical Assistance. CWA – Clean Water Act. CZARA – Coastal Zone Act Reauthorisation Amendments. EQIP – Environmental Quality Improvement Program. ESA – Endangered Species Act. EWRP – Emergency Wetland Reserve Program. FIFRA – Federal Insecticide, Fungicide and Rodenticide Act. WHIP – Wildlife Habitat Incentives Program. WQIP – Water Quality Improvement Program. WRP – Wetland Reserve Program.
Source: Heimlich and Claassen (1998), p. 98.

Figures 6.7 and 6.8 compare historical technical assistance costs for US land retirement and cost-sharing programmes between 1983 and 2002. Technical assistance costs for land retirement programmes in the US have typically run at 5 to 10% of expenditures for rental/easement and cover establishment cost-sharing. However, there were large start-up costs (110%) in the first year of both the CRP and WRP programs. The Water Bank (WBP) was an older land retirement programme started in the 1950s that paid farmers to retain shallow wetlands and buffers in farming areas. Water Bank was merged with WRP after the 1990 FACTA Act, so the abrupt change in technical assistance percentage (from 0.1% to 4.5%) is more likely due to a change in accounting for technical assistance than any actual change in the amount of assistance provided. There may have been some additional transaction costs as existing Water Bank agreements were converted to easements under WRP.

Figure 6.7. **Land retirement programmes: Technical assistance as a per cent of cost-share and rental/easement expenditures**

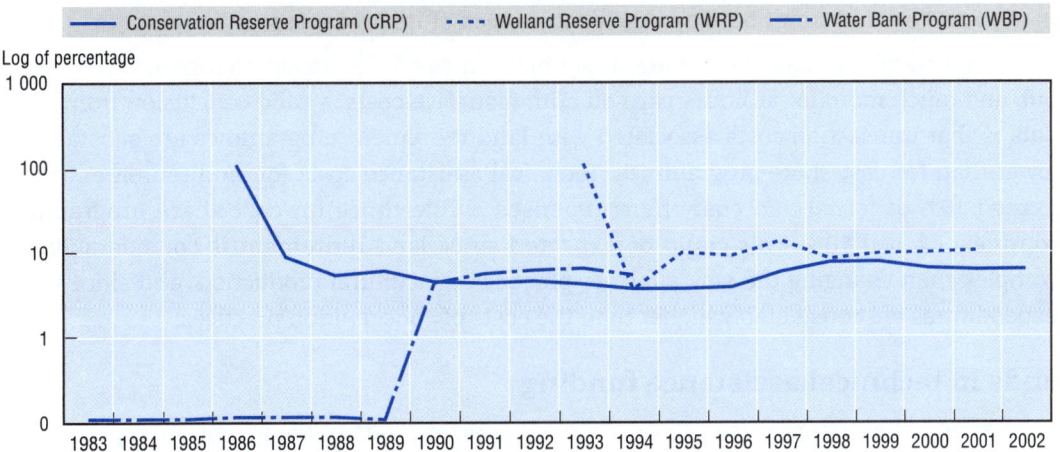

Source: USDA, OBPA and Ralph E. Heimlich, Agricultural Conservation Economics.

Figure 6.8. **Cost-share programmes: Technical assistance as a per cent of cost-share expenditures**

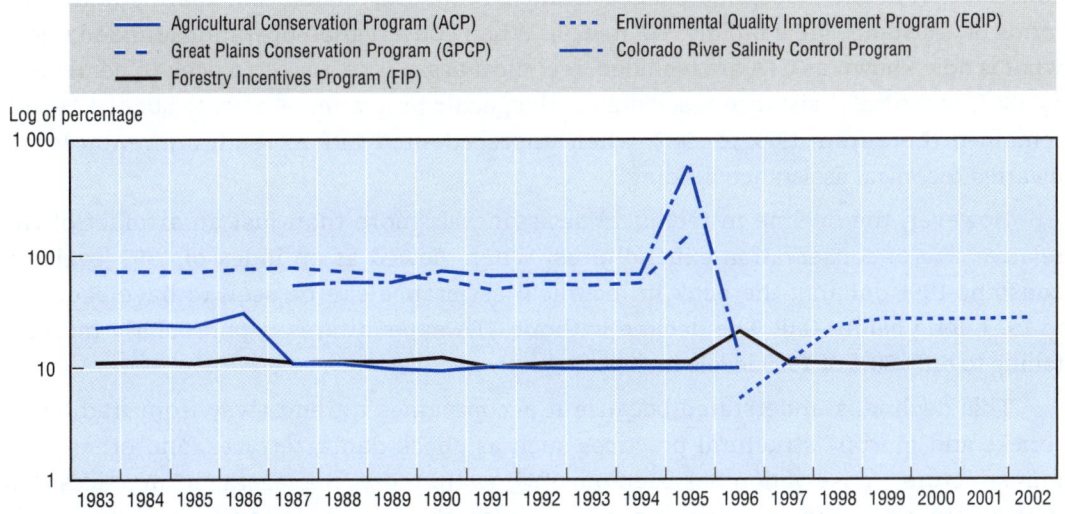

Source: USDA, OBPA and Ralph E. Heimlich, Agricultural Conservation Economics.

Both GPCP and the CRBSCP had much higher technical assistance costs than the other cost-share programmes and than land retirement programmes, ranging from 55 to 80% of cost-share expenditures. This is less inefficiency than design, since these programmes focused more on conservation management changes and less on capital investments or "hardware" expenditures.

Much of the technical assistance work supporting ACP was probably funded under the NRCS CTA budget, which focused on continuing planning with long-term relationships (see whole farm planning discussion). The abrupt change in the ACP technical assistance percentage (from 31 to 11%) after the 1985 Farm Act probably reflects administrative accounting decisions rather than real changes in technical assistance available to producers, particularly since the CTA budget increased substantially at this time (from USD 410 to USD 458 million). The surges in technical assistance percentage for GPCP, CRBSCP, and FIP in 1996 reflect shut-down and transitional expenditures of consolidating those programmes and developing EQIP, which ramped up from 1996 to 1998, then settled in at 26-27%, a mid-point between the percentages of its predecessors.

While there is substantial variation as programmes are developed, change and are phased out, and much room for arbitrary caps on administrative costs, a valid conclusion from these data is that transaction costs associated with land retirement programmes are substantially lower than for cost-share programmes. Technical assistance costs for land retirement rarely exceed 10% of rental and cost-share expenses, while those for cost-share programmes routinely exceed 10%. This could be expected since land retirement is considerably less complex than changing the way a farmer practices agricultural production, and since rental expenditures are considerably higher than the costs of conservation practices.

6.6. Trends in technical assistance funding

Technical assistance used to be a more important part of conservation expenditures than it is today. Congress increased conservation expenditures in the 1996 and 2002 omnibus farm legislation, but did not maintain the proportion of those expenditures earmarked for helping farmers use the increased funding effectively to solve conservation problems. Although technical assistance was nearly all the help farmers got for conservation in the 1930s, financial assistance, rental payments, cost-sharing and other forms of assistance grew rapidly (Figure 6.9). When conservation operations funded under what is now known as CTA are included, technical assistance grew to a peak of nearly 70% by 1987. Technical assistance associated with specific programmes grew to about 17% and remained there from 1972 to 1987, when unprecedented CRP expenditure rental levels dwarfed technical assistance funding.

However, the decline in technical assistance is more than just an artefact of the relative size of conservation expenditures. When viewed as an index of 1999 levels (in constant 1996 dollars), the peak in technical assistance can be seen to have occurred in 1973, long before CRP. The decline without CTA expenditures is particularly marked, falling to only 30% of 1999 levels in 1998.

This decline is understated because it accompanies a trend away from traditional "bricks and mortar" structural practices such as check dams, terraces, and other built infrastructure. Conservation has been moving toward more "management" practices that involve changing the way farming is done that should require even more technical assistance time with farmers.

Figure 6.9. **Technical assistance as a per cent of conservation expenditures 1937-99**

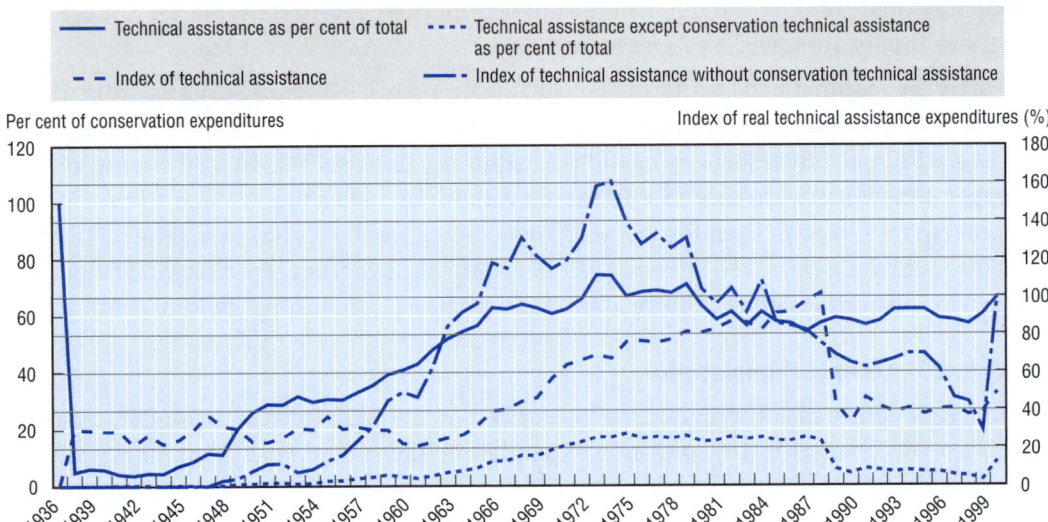

Source: USDA, ERS AREI Chapter 6.1 and Ralph E. Heimlich, Agricultural Conservation Economics.

Four provisions in the 2002 FSRI Act highlight congressional technical assistance funding: the Section 11 funding limitation, limits on technical assistance for the Conservation Security Program (CSP), passage of the Technical Service Provider (TSP) option, and reform of conservation planning. In 2002, the Federal Office of Management and Budget issued its A-70 regulations and both the Executive and Legislative Branches began turning serious attention to reducing the public costs of doing government business. Thus, the FSRI Act reflects trends outside of agriculture toward reducing government expenditures and increasing efficiency.

Section 11 cap

Just as funding for CRP rental and cost-share payments, funding for NRCS technical assistance for CRP is done under the Commodity Credit Corporation (CCC) authorisation, rather than through direct appropriations. CCC is a federal corporation that is located within the USDA and managed by a Board of Directors under the supervision of the Secretary of Agriculture. CCC is empowered to obtain funds through borrowing, as well as through direct appropriations from Congress. Section 11 of the CCC Charter Act authorizes the CCC to allot or transfer "to any bureau, office, administration or other agency of the Department of Agriculture [...] any of the funds available to [the CCC] for administrative expenses", 15 USC. § 714i. Originally founded to fund commodity programmes, which often involved receipts for sales of surplus commodities, CCC has been increasingly tapped for other kinds of programmes, especially conservation programmes, since the 1996 Farm Bill.

A provision intended to limit unauthorized use of CCC funding authority for data processing and information technology purchases put a binding limit on reimbursable agreements between Federal agencies that may have unintentionally limited use of CCC funds by NRCS for technical assistance to the total amount of the allotments and transfers in fiscal year 1995, or about USD 56 million. Section 161, H.R. 2854, P.L. 104-127.

USDA was forced to pay for some CRP technical assistance out of funds appropriated for the CTA program for ongoing conservation operations outside the farm bill programmes. Various legislative fixes for the Section 11 cap failed to pass, causing a bureaucratic disagreement about where CRP technical assistance should be funded until S.2856 was finally approved on 7 December 2004.

Partly in response to limits on NRCS funding for CRP technical assistance, FSA implemented a number of administrative measures to improve programme delivery while reducing administrative delivery costs. For example, during CRP general signup 26, FSA developed a new software tool to automate evaluations using the EBI and to provide GIS support in many counties. Over the last year, this GIS tool greatly reduced the time required for farmers to submit offers, saved farmers USD 160 000 in participation expenses, and helped FSA reduce administrative costs for CRP by over USD 7 million (Little testimony, May 2004).

Limit on CSP technical assistance

In a new conservation programme authorized in the 2002 Farm Bill, the CSP, technical assistance was limited to 15% of funds expended in each fiscal year. Senator Harkin, the programme's sponsor, in responding to delays in implementing the new programme said that the cap on technical assistance should not impede implementation because technical assistance will require a much lower percentage of total CSP funds since CSP emphasizes maintaining less expensive, already existing practices and precludes very expensive manure transport and storage practices with high technical assistance costs (Harkin, 2004).

This observation fails to account for several aspects of modern conservation programmes. A large part of technical assistance costs are spent on simply ensuring that producers are eligible for the programme. Assisting farmers navigate the increasingly complex requirements of recent conservation programmes takes considerable time, technical investigation, and face-to-face interaction with producers. Second, the number of applicants is often large relative to the number ultimately enrolled in the programme. Thus, a portion of technical assistance monies are expended on would-be applicants who are not ultimately successful and do not end up applying conservation practices. Third, modern conservation requires use of "management" practices that require considerable interaction with a conservation professional to assist the producer develop a nutrient management plan, select less environmentally damaging pesticides that are equally effective in controlling pest pressures, or assess range or pasture degradation and devise restoration strategies. Much of the older engineering assistance is actually encompassed in the cost-share financial assistance because it is done by commercial firms that implement the structures or construction embodied in those practices.

Initial implementation of CSP in July 2004 relied heavily on an online self-assessment filled out by the producer, reviewed by NRCS technical service providers, but with limited on-farm evaluation and planning (USDA, NRCS, 2004). NRCS Chief Bruce Knight is initially highly encouraged by this approach, and sees it as a model for future programme delivery (Rainford, 2004). While this use of information technology may constitute an increase in efficiency over paper forms and face-to-face interviews, it heavily substitutes untrained producer input for on-the-ground assessment by trained conservation professionals that may prove counterproductive in the longer term.

Third-party technical assistance

The 2002 FSRI Act also established a private-sector alternative to government provided technical assistance in the form of third-party technical service providers (Section 2701, FSRI Act of 2002, amending Section 1242 of the FSA Act of 1985; Federal Register Vol. 67,

No. 225, Thursday, 21 November 2002, 70119:70133). The Senate version of the legislation required USDA to establish provisions for increased technical assistance by non-federal providers, including certification of providers and cooperative agreements with state, local and nongovernmental groups to provide technical assistance. Congress recognized that, while USDA had been the primary provider of technical assistance to conservation programme participants, it would be difficult to meet the increased demand for technical services under the increased financial assistance over the life of the farm bill. The potential volume of many new, as well as returning, USDA conservation programme participants would overwhelm the assistance available through existing USDA resources, but Congress was reluctant to increase governmental staff levels. To meet this demand, Congress encouraged assistance from third-party providers, combining both the private and public sectors to provide technical assistance for USDA conservation programmes.

The Managers intended that third-party vendors accepting federal technical assistance payments follow all the applicable Federal laws and accept the appropriate liability for the adequacy of their plans, practice designs, and implementation procedures, and to comply with all appropriate privacy and confidentiality requirements. Putting these burdens, long shouldered by government technical assistance providers, on TSPs may make it less likely that private-sector providers can operate more efficiently than government providers.

Conservation planning reform

A provision of the 2002 FSRI Act also expressed dissatisfaction with the fragmented nature of conservation planning and technical assistance provision by calling on USDA to develop a plan to coordinate land retirement and agricultural working land conservation planning to eliminate redundancy, streamline programme delivery, and improve services provided to agricultural producers, including re-evaluation of the provision of technical assistance. A report is required by the end of 2005 to present a plan to integrate conservation planning programmes and the means to implement the plan.

These four provisions show different aspects of Congressional efforts to provide for conservation technical assistance. Technical assistance in planning and evaluating applicants for limited conservation assistance is necessary for cost-effective implementation. While technical assistance has some elements of a policy-related transaction cost that programme managers should strive to minimize, it is also an essential component in selecting the best applicants to receive conservation assistance, and helping those applicants develop and implement new ways of farming that will help conserve their resources and the environment.

6.7. Conclusions

Overall, the costs to the government of implementing the Conservation Reserve Program are relatively low, running from 3% of expenditures in initial years and 1% in succeeding years for NRCS technical assistance, and about 4% of expenditures for FSA administrative support costs. This amounts to about USD 60 per acre enrolled in initial years of a 10-year enrolment period, and about USD 20 per acre in succeeding years. These costs are less than comparable costs for the Wetland Reserve Program, and much less for working land programmes such as EQIP and its predecessor programmes. The absolute size of rental payments in CRP dwarfs transaction costs in ways that cost-share funds under working lands programmes do not.

FSA administrative costs are highly correlated with programme characteristics, especially the cumulative acreage enrolled, with each additional acre increasing costs by USD 1.79. NRCS technical assistance costs are more variable, and are significantly correlated with acres idled or installed in a given year (adding USD 2.39 per acre) and cumulative acres enrolled (adding USD 0.30 with each additional acre) in each year. NRCS costs increased significantly between the first CRP signups and the second set after 1996. The signs and magnitudes of other correlates are interesting, but not statistically significant.

Conservation technical assistance overall has declined from peak levels in the mid-1970s, despite an increase in management-related practices. Congressional support for technical assistance may be dropping, as evidenced by issues related to the Section 11 cap on reimbursement, caps on technical assistance in the new CSP, reliance on third-party technical assistance providers, and mandates for studying conservation planning reform.

Information technology, centralisation of functions, and other administrative improvements can reduce technical assistance and administrative transaction costs, and improve the ability to evaluate resource concerns and conservation plans to correct them, and has partly compensated for reduced technical assistance funding available in recent years. However, continued decreases in technical assistance at the field level and reliance on online resources and information technology cannot indefinitely substitute for face-to-face, on-the-ground technical assistance provided by trained conservationists to producers interested in learning about and applying improved methods. Technical assistance is not merely a cost or friction to be overcome for more efficient programme implementation, but part of the programme itself.

Notes

1. For example, in the context of stock markets, transaction costs include the time, effort, and money necessary, including such things as commission fees and the cost of physically moving the asset from seller to buyer. Transaction costs should also include the bid/ask spread as well as price impact costs (for example a large sell order could lower the price). See *www.marketvolume.com/glossary/t0282.asp*

2. Prior to the 1994 USDA reorganisation, NRCS was known as the Soil Conservation Service (SCS) and FSA was the Agricultural Conservation and Stabilization Service (ASCS). These names dated from the agencies' establishment in the 1930's.

References

Allen, A.W. (1994), *Regional and State Perspectives on Conservation Reserve Program Contributions to Wildlife Habitat*. US Fish and Wildlife Service Federal Aid Report, National Ecology Research Center, Fort Collins, CO, 28 p.

American Agricultural Economics Association (1986), *Soil Erosion and Soil Conservation Policy in the United States*, Occasional Paper No. 2, AAEA Soil Conservation Policy Task Force, January.

Barbarika, A., C.T. Osborn and R.E. Heimlich (1994), "Using an Environmental Benefits Index in the Conservation Reserve Program," in *Proceedings of the NCT-163 Post Conservation Reserve Program Land Use Conference*, Denver, CO, January 10-11, pp. 118-133.

Barlow, C.P. (1989), "Stress in the Soil Conservation Service", *Journal of Soil and Water Conservation* 44(2), pp. 105-110, March-April.

Berner, A.H. (1989) "The 1985 Farm Act and its Implications for Wildlife," in W. Chandler (ed.) *Conservation Challenges, Audubon Wildlife Report*, 1988/89, Academic Press, Inc. pp. 436-465.

Bills, N.L. and R.E. Heimlich (1984), *Assessing Erosion on US Cropland: Land Management and Physical Features*, AER-513, US Departement of Agriculture, Economic Research Service, July.

Bridge, G. (1993), "Is whole-farm Conservation Planning the Answer?", *Journal of Soil and Water Conservation* 48(4):295-298, July-August.

Christensen, R.P. and R.O. Aines, (1962), *Economic Effects of Acreage Control Programs in the 1950s*, AER-18, US Departement of Agriculture, Economic Research Service, October.

Claassen, R., L. Hansen, M. Peters, V. Breneman, M. Weinberg, A. Cattaneo, P. Feather, D. Gadsby, D. Hellerstein, J. Hopkins, P. Johnston, M. Morehart and M. Smith (2001), *Agri-environmental Policy at the Crossroads: Guideposts on a Changing Landscape*. AER-794, US 2001, January.

Clark, A., M. Havercamp and W. Chapman (1985), *Eroding Soils: The Off-farm Impacts*, Washington DC, The Conservation Foundation.

Cohee, M.H. (1986), "The Soil Conservation Imperative: Past *versus* Present", Journal of Soil and Water Conservation 41(2):94-96, March-April.

Council for Agricultural Science and Technology (1990), *Ecological Impacts of Federal Conservation and Cropland Reduction Programs*, Report No. 117, Ames, IA, September.

Crosson, P. and A.T. Stout. (1983), *Productivity Effects of Cropland Erosion in the United States*. Washington DC, Resources for the Future.

Conference Report on H.R. 2646, Farm Security And Rural Investment Act of 2002 (House of Representatives – 1 May 2002), online at *http://thomas.loc.gov/cgi-bin/query/F?r107:1:./temp/~r107MkCiPm:b1491646*, last accessed 8/11/2004.

Dicks, M.R. (1985), "Aggregate Economic Impacts of a Conservation Easement Program for the Corn Belt", unpublished dissertation, University of Missouri-Columbia.

Ervin, D.E. and J.W. Mill (1985), "Agricultural Land Markets and Soil Erosion: Policy Relevance and Conceptual Issues", *American Journal of Agricultural Economics* 67(5): 938-42, 1985.

Ervin, D.E. and M.G. Blase (1986), "The Conservation Reserve: Potential Impacts and Problems", *Journal of Soil and Water Conservation* 41(2):77-80, March-April.

Farnsworth, R.L. and J.B. Braden (1988), "Educational and institutional needs of the Conservation Title", *Journal of Soil and Water Conservation* 44(5), pp. 395-398, September-October.

Farnsworth, R.L., R.J. Herman and R.D. Walker (1988), "Workshops for Integrating Resource Management and Agricultural Production", *Journal of Soil and Water Conservation* 44(5), pp 399-402, September-October.

Feather, P., D. Hellerstein and L. Hansen (1999), *Economic Valuation of Environmental Benefits and the Targeting of Conservation Programs: The Case of the CRP*, US Department of Agriculture, Economic Reserach Service, AER-778, April.

Gray, R.J. (1986), "Proving Out: On Implementing the Conservation Title of the 1985 Farm Bill", *Journal of Soil and Water Conservation* 41(1), pp. 31-32, January-February.

Harkin, T. (2004) "Harkin: USDA is Missing the Boat with CSP," US Senate, Tuesday, 11 May 2004, online at *www.harkin.senate.gov/news.cfm?id=221414*, last accessed 8/11/2004.

Hawn, T. and M. Getman (1992), "Enhancing CRP Values", *Journal of Soil and Water Conservation* 47(2), pp. 134-135, March-April.

Heimlich, R.E. (2002), "The US Experience with Land Retirement for Natural Resource Conservation," and "Evaluating Bids in the US Conservation Reserve Program," Xu Jintao and Ulrich Schmitt (Eds.), Workshop on Payment Schemes for Environmental Services: Proceedings, CCICED Task Force on Forests and Grasslands, Beijing, 22-23 April, China Forestry Publishing House, pp. 12-15 and 36-38 (full text on CD-ROM).

Heimlich, R.E. and R.C. Claassen (1998), "Agricultural Conservation Policy at a Crossroads", *Agricultural and Resource Economics Review*, Vol. 27, No. 1, April, pp. 95-107.

Larson, W.E., F.J. Pierce and R.H. Dowdy (1983), "The Threat of Soil Erosion to Long-term Crop Production", *Science*, 219(4584):458-465.

Letter for S.A. Poling (2002), Associate General Counsel, General Accounting Office, from P.J. Perry, General Counsel, Office of Management and Budget (16 September 2002).

Letter for S.A. Poling (2002), Associate General Counsel, General Accounting Office, from N.S. Bryson, General Counsel, US Department of Agriculture (16 September 2002).

Letter for Senator T. Harkin (2002), Chairman, Senate Comm. on Agriculture, Nutrition and Forestry, from N.S. Bryson, General Counsel, US Department of Agriculture (24 September 2002), quoting electronic message communicating the Congressional Budget Office's conclusion that the Section 11 ceiling remains "applicable to the transfers under Section 1241(a)".

Letter for Senator H. Kohl (2002), Chairman, Subcommittee on Agriculture, Rural Development, and Related Agencies, Senate Appropriations Comm., Senator T. Cochran, Ranking Minority Member, Subcommittee on Agriculture, Rural Development, and Related Agencies, Senate Appropriations Comm., and Representative H. Bonilla, Chairman, Subcomm. on Agriculture, Rural Development, FDA & Related Agencies, House Appropriations Comm., from A.H. Gamboa, General Counsel, US General Accounting Office, *Re: Funding for Technical Assistance for Conservation Programs Enumerated in Section 2701 of the Farm Bill*, No. B-291241 (8 October 2002), available at *www.gao.gov*).

Little, James Administrator, Farm Service Agency (2004), Testimony before the Forestry, Conservation, and Rural Revitalisation Subcommittee, Senate Agriculture, Nutrition and Forestry Committee "Examining Conservation Programs of the 2002 Farm Bill", 11 May 2004, SD-628, Dirksen Senate Building online at *http://agriculture.senate.gov/Hearings/hearings.cfm?hearingId=1163*, last accessed 8/11/2004.

Naley, M. (2004), US Department of the Interior, Fish and Wildlife Service, Habitat Restoration Branch, personal communication, 8/3/2004.

National Association of Conservation Districts (2001), Technical Assistance – The Key to Conservation Improvements on America's Working Lands (28 August), online at *www.nacdnet.org/govtaff/FB/TA-theKey.htm*, last accessed 8/11/2004.

Nielson, J. (1986), "Conservation Targeting: Success or failure?" *Journal of Soil and Water Conservation* 41(2), pp. 70-76, March-April.

Nowak, P. and M. Schnepf (1988), "Implementing the Conservation Provisions in the 1985 Farm Bill: A Survey of County-level US Department of Agricultural Agency Personnel", *Journal of Soil and Water Conservation* 42(4), pp. 285-290, July-August.

Nowak, P. and M. Schnepf (1989), "Implementing the Conservation Provisions in the 1985 Farm Bill: A Follow-up Survey of County-level US Department of Agricultural Agency Personnel", *Journal of Soil and Water Conservation* 44(5), pp. 535-541, September-October.

Ogg, C.W., M.P. Aillery and M.O. Ribaudo (1989), *Implementing the Conservation Reserve Program: Analysis of Environmental Options*, AER-618, US Dept. Agr., Econ. Res. Serv., October.

Osborn, C.T., F. Llacuna and M. Linsenbigler (1995), *The Conservation Reserve Program: Enrolment Statistics for Signup Periods 1-12 and Fiscal Years 1986-93*, SB-925, US Department of Agriculture, Economic Research Service, November.

Rainford, C. (2004), "NRCS Chief Optimistic CSP Will Be a Vibrant, Rapidly Growing Program", *Agriculture Online*, 4 May, online at *www.agriculture.com/default.sph/AgNews.class?FNC=topStoryDetail__ANewsindex_html__51714__1*, last accessed 8/13/2004.

Reichelderfer, K.H. and W.G. Boggess (1988), "Government Decision-making and Program Performance: The Case of the Conservation Reserve Program", *American Journal of Agricultural Economics* 70(1), pp. 1-11.

Rey, M. (2004), Statement of Under Secretary, Natural Resources and Environment, United States Department of Agriculture, Before the House Appropriations Subcommittee on Agriculture, Rural Development, Food and Drug Administration, and Related Agencies, 26 February, online at *http:// appropriations.house.gov/index.cfm?Fuseaction=Hearings.Testimony&HearingID=314&WitnessID=523*, last accessed 8/11/2004.

Ribaudo, M.O (1986), *Reducing Soil Erosion: Offsite Benefits*, AER-561, US Department of Agriculture, Economic Research Service, September.

Robertson, T., G. Root and K. Reinhardt (1989), "Conservation Planning: Group *Versus* Individual Approaches", *Journal of Soil and Water Conservation* 44(5):395-398, September-October.

Taylor, M.R. (2001), "The Emerging Merger of Agricultural and Environmental Policy: Building a New Vision for the Future of American Agriculture", *Virginia Environmental Law Journal*, 20(1):169-190.

US Department of Agriculture (1997), Farm Service Agency, *Fact Sheet, Environmental Benefits Index, Conservation Reserve Program Sign-up*, 16 October 1997, online at *www.fsa.usda.gov/pas/publications/ facts/crp16ebi.pdf*.

US Department of Agriculture, Economic Research Service (2000), Overview of Conservation Programs and Expenditures, Chapter 6.1, *Agricultural Resources and Environmental Indicators*, AHB-722, online at *www.ers.usda.gov/publications/arei/ah722/arei6_1/DBGen.htm*, last accessed 8/13/2004.

US Department of Agriculture, Farm Service Agency (2003), Conservation Reserve Program Final Programmatic Environmental Impact Statement, January, online at *www.fsa.usda.gov/dafp/cepd/ epb/impact.htm#final*.

US Department of Agriculture, Farm Service Agency (2003), *Fact Sheet, Environmental Benefits Index, Conservation Reserve Program Sign-up*, 26 May, online at *http://www.fsa.usda.gov/pas/publications/ facts/html/crpebi03.htm*.

US Department of Agriculture, Farm Service Agency (2004), Conservation Reserve, *Monthly Summary*, June 2004, online at *www.fsa.usda.gov/dafp/cepd/stats/JUN2004.pdf*, last accessed 8/13/2004.

US Department of Agriculture, Farm Service Agency (2004b), Conservation Reserve Enhancement Program National Summary, January, online at *www.fsa.usda.gov/dafp/cepd/crep/summary.htm*, last accessed 8/13/2004.

US Department of Agriculture, Natural Resources Conservation Service (2004), Conservation Security Program: Self-Assessment Workbook, PA-1770, June, online at *http://www.nrcs.usda.gov/programs/ csp/pdf_files/CSP_SelfAssess_Workbook_F.pdf*.

US Department of Agriculture, Office of Budget and Program Analysis (2002), USDA Conservation Funding by Agency and Program, 1983-2002 (unpublished tables).

US Department of Agriculture, Soil Conservation Service (1994), *1992 National Resources Inventory: Highlights*, EI&D-94-920, July.

US Department of Justice, Office of Legal Counsel (2003), *Funding For Technical Assistance For Agricultural Conservation Programs, Memorandum For The General Counsel*, Office Of Management and Budget, 3 January, online at *http://www.usdoj.gov/olc/usdasection11.htm*.

United States General Accounting Office (2000), USDA Reorganisation: Progress Mixed in Modernising the Delivery of Services Report to the Chairman, Committee on Agriculture, Nutrition, And Forestry, US Senate, GAO/RCED-00-43, February, online at *www.gao.gov/archive/2000/rc00043.pdf*.

Ventura, S.J. and D.A. Giampetroni (1992), "Wisconsin Conservationists Respond to Field Office Overload", *Journal of Soil and Water Conservation* 48(2), pp. 83-89, March-April.

OECD PUBLICATIONS, 2, rue André-Pascal, 75775 PARIS CEDEX 16
PRINTED IN FRANCE
(51 2007 01 1 P) ISBN 978-92-64-03091-6 – No. 55495 2007